Iosias Jody (Ed.)

Louisville and Nashville Railroad Office Building

Iosias Jody (Ed.)

Louisville and Nashville Railroad Office Building

Downtown Louisville, Kentucky, Louisville and Nashville Railroad

Cred Press

Contents

Louisville and Nashville Railroad Office Building

Louisville and Nashville Railroad Office Building	
U.S. National Register of Historic Places	
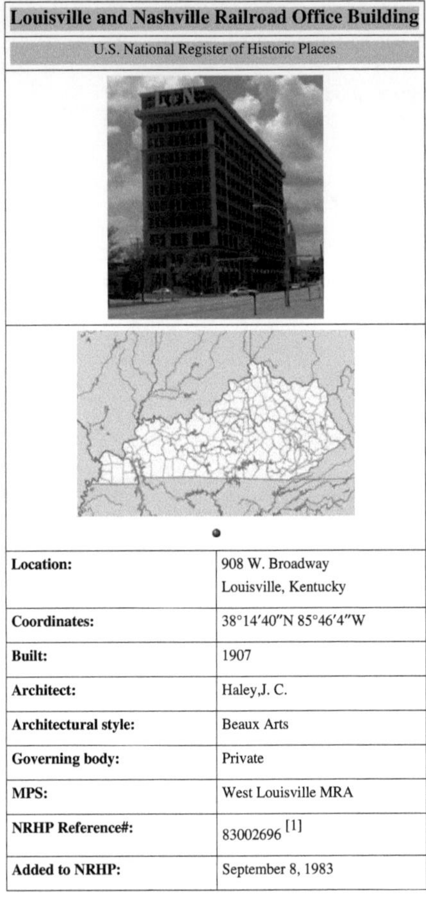	
Location:	908 W. Broadway Louisville, Kentucky
Coordinates:	38°14′40″N 85°46′4″W
Built:	1907
Architect:	Haley,J. C.
Architectural style:	Beaux Arts
Governing body:	Private
MPS:	West Louisville MRA
NRHP Reference#:	83002696 [1]
Added to NRHP:	September 8, 1983

The **Louisville and Nashville Railroad Office Building** is an historic building located in downtown Louisville, Kentucky, USA. It was once the headquarters of the Louisville and Nashville Railroad, a prominent railroad company from the mid-19th century to the 1970s.

Architecture

The structure is eleven stories tall. The first three stories are made of stonework of rusticated ashlar, with capital-topped pilasters in a series. Floors four to ten have ashlar pilasters framing a finish of red brick. Windows of the building are done in series of three. The attic is 1.5 stories tall, and features the distinctive initials of the Louisville and Nashville Railroad.[2]

It was designed by W. H. Courtenay, the chief architect of the Louisville and Nashville Railroad, in a Beaux Arts style; one of the largest commercial buildings in that architectural style still standing.[3] [4]

In 1930 an eight-bay western addition which duplicated the look of the original 10-story building was added, built by then-current chief architect of the Louisville and Nashville Railroad J. C. Haley.[5]

History

Aerial view of the L&N freight (left) and passenger (right) stations in the first half of the 20th century, showing the L&N Office Building and Union Station.

The original Louisville and Nashville Railroad offices in Louisville were at Second and Main in Louisville, by the entrance of present-day George Rogers Clark Memorial Bridge. By 1890, it had become obvious that the building was too overcrowded. It was decided that the office building should be located next to Louisville's Union Station. Construction began in 1902, but its completion was delayed until January 1907, due to difficulties with organized labor in a 1905 steel workers strike. Its total cost was $650,000. It was large enough that after decades of separation, all of the main administrative staff could be in the same building.[6]

In the 1970s, about 2,000 L&N employees worked in the building. After L&N was purchased by CSX nearly all of the jobs were moved from Louisville to Jacksonville, Florida, in 1980. However, a 6-person CSX claims department stayed in the building until 1988.[7]

In 1984, the state of Kentucky spent $15 million to purchase and renovate the property, retaining the L&N name and neon lights on its upper stories.[7]

The building is currently the Louisville offices for the Kentucky Cabinet for Health and Family Services.[8]

In August 2009, the building was closed due to the 2009 Kentuckiana Flood, but would reopen as soon as deemed safe.[9]

References

[1] "National Register Information System" (http://nrhp.focus.nps.gov/natreg/docs/All_Data.html). *National Register of Historic Places*. National Park Service. 2008-04-15. .

[2] Luhan pg.74, 75

[3] Luhan, Gregory. Louisville Guide. (Princeton Architectural Press, 2004) pg. 74

[4] Klein, Maury. *History of the Louisville & Nashville Railroad*. (University Press of Kentucky, 2002). pg.340

[5] Luhan pg.74

[6] Klein, Maury. *History of the Louisville & Nashville Railroad*. (University Press of Kentucky, 2002). pg.340

[7] Hershberg, Ben (1988-11-05). "LAST OF L&N WORKERS ARE LEAVING LANDMARK". Courier-Journal. pp. B12.

[8] "Resource Directory" (http://www.louisvillerelationships.org/ResourceDirectory.htm). Resource Education Across Louisville. . Retrieved 2008-07-01.

[9] Yetter, Deborah (August 5, 2009). "L&N building offices closed by flooding" (http://www.courier-journal.com/article/20090805/NEWS01/908050383/). . Retrieved 2009-08-05.

Downtown Louisville

Downtown Louisville is the largest central business district in the Commonwealth of Kentucky and the urban hub of the Louisville, Kentucky Metropolitan Area. Its boundaries are the Ohio River to the north, Hancock Street to the east, York and Jacob Streets to the south, and 9th Street to the west. As of 2000, the population of Downtown Louisville was 2,575.

The five main areas of the Central Business District consist of:

The Louisville Skyline

- **West Main District** (west of 2nd St., north of Market St., east of 9th St., and south of the Ohio River)
- **East Main District** (east of 2nd St., north of Market St., west of Hancock St., and south of the Ohio River; contains the Whiskey Row Historic District)
- **Medical Center** (east of 2nd St., south of Market St., west of Hancock St., and north of Jacob St.)
- **Fourth St. District** (south of Market St., west of 2nd St., north of York St., and east of 5th St.)
- **Civic Center** (south of Market St., west of 5th St., north of York St., and east of 9th St.)

The tallest buildings in Kentucky are located in Downtown Louisville and include the AEGON Center designed by John Burgee, National City Tower designed by Harrison & Abramovitz, PNC Plaza designed by Welton Becket, and the Humana Building designed by Michael Graves. Of the 16 buildings in Kentucky over 300 feet (91 m), 12 are in Downtown Louisville. In addition, it is the center of local and regional government.

A glassed-in skywalk called the Louie Link stretches six city blocks and links together the Kentucky International Convention Center (KICC), Fourth Street Live!, three hotels (Galt House Hotel & Suites, Marriott and Hyatt Regency), and 2,300 hotel rooms. In 2010 it was extended from the Galt House to the new $16 million Skywalk Garage, an eight-level, 860-space parking facility on Third Street, and a second skywalk connects from the garage across Third Street to the new KFC Yum! Center.

The AEGON Center

History

Downtown Louisville is the oldest part of the city of Louisville, whose initial development was closely tied to the Ohio River. The largest early fort, Fort Nelson, was built in 1781 near what is today the corner of 7th and Main streets. Many early residents lived nearby after moving out of the forts by the mid-1780s, although little remains from of the earliest (mostly wood) structures.

Early plans of the city, such as William Pope's original plan in 1783, show a simple grid on an east/west axis along the river. The earliest streets, Main, Market and Jefferson retain their original names from the plan, while the smaller Green Street is now known as Liberty (it was renamed after Green Street acquired a seedy reputation due to its many burlesque theaters). Main Street was the city's initial commercial hub for nearly a century.

By 1830 Louisville passed Lexington as Kentucky's largest city, with a population over 10,000. The steamboat era saw the opening of the Louisville and Portland Canal just west of downtown, and local commerce picked up further with the founding of banks and manufacturing. Most of Louisville's population was packed into downtown, which by this time stretched as far south as Prather Street (later renamed Broadway). Many still-remaining buildings reveal what the area was like at this time, with narrow, two to four-story buildings packing the streets.

The Richardsonian Romanesque Levy Building, built in 1893, is an example of Downtown Louisville's classic architecture, as well as its revitalization; the upper four floors have been converted to 23 loft condominiums.

The area and the city continued to grow during the railroad era. However, the increased mobility of early trolleys, as well as the shear number and diversity of people moving to Louisville, saw a shift in focus as areas like Phoenix Hill, Russell and what is now Old Louisville began to be built on the edges of downtown, particularly after the city annexed those areas in 1868. Railroads lead to a diminished role for the river in transportation, further reducing the importance of downtown in favor of areas on what was then the edge of the city, along rail lines.

In 1886, the first skyscraper,[1] the Kenyon Building, was completed on Fifth Street, followed in 1890 by the ten-story Columbia Building. The development of three large suburban parks and the electrified streetcar lead to the first true movement to the suburbs at this time. Some of downtown's business and industry followed people toward these areas. But by the 1920s the commercial center of Louisville was still nearby, at 4th and Broadway, dubbed the "magic corner" by the *Herald-Post*. The riverfront area of downtown was still being actively improved, such as with the building of what is now George Rogers Clark Memorial Bridge across the Ohio at Second Street in 1929.

After World War II, suburbanization increased and downtown began to decline as interstate highways further reduced the importance of its central location. Since the 1970s, downtown has been the subject of both urban renewal and historic preservation efforts. While many new buildings have been built, it has sometimes been at the expense of older landmarks, such as the Tyler Block.

Many buildings sat totally or mostly vacant at this time, and some became dilapidated to the point where they burned down or had to be razed. Many riverfront industrial sites were abandoned or saw limited use, many were eventually redeveloped into Louisville Waterfront Park. Other issues in the 1970s through the early 1990s included a former theater district on Jefferson Street that had become dubbed the "porno district". The businesses there were seen by the city as an eyesore since they were so close to the convention center, and most were demolished or burned down by the late 1990s. A few adult book stores and bars remained in the general area as of 2007.

From the late 1970s to early 1990s, nine new high rises over 200 feet (61 m) in height were built in downtown. Unlike the city's previous tallest buildings, which were all set along the Broadway corridor, these new buildings were set closer to the riverfront along Main and Market Streets.

Since 2000, downtown has seen another major growth spurt, although this one not only includes new high rises, but also a large scale return of large scale residential and retail back to the city center. The completion of Louisville Slugger Field along with a mass expansion of the city's Waterfront Park, both completed in 1998, sparked new development along the eastern edge of downtown, with entire abandoned blocks rebuilt with new condominium units and shops. Also, new to Louisville is the 22,000-seat KFC Yum! Center at Second and Main Streets which was completed in 2010.

The Columbia Building, Louisville's second skyscraper

The Brown Hotel (built 1923)

The Heyburn Building (built 1928)

Residential

Early residences outside of the forts, still mostly wood structures, were built along the modern street grid on early lots sold to settlers, but have all been demolished over time. What became the almost entirely office and parking-lot dominated downtown still had many solidly single family residential blocks on its fringes up until the early 20th century. Streets near Broadway, such as Chestnut, were lined with large mansions of the owners of businesses on Main and Market streets.

Example of an early downtown mansion, this Walnut Street mansion was built by the Belknap family and used as the first clubhouse of the Pendennis Club before being razed

New development on East Main Street

Though these houses were built of brick and other longer-lasting materials, none remained single family homes by the 21st century, although some had been converted for other uses, such as office space. The Brennan House at 631 S. Fifth, which is operated as a historic property with daily tours, shows a glimpse of Downtown Louisville's residential past. A structure at 432 South Fifth Street is the only example of a pre-Civil War residence remaining Downtown; built in 1829 it has been converted to commercial use.[2]

By the late 20th century, downtown Louisville had acquired a reputation as a place to work and visit during the week but which shuts down evenings and weekends. The first changes to this were the conversion of old warehouse and factory space to loft apartments in the late 1980s. By the late 1990s and early 2000s, new developments of luxury condominiums such as the 22-story *Waterfront Park Place*, and the $30 million project *Fleur de Lis on Main*, indicate increasing residential interest in Downtown Louisville. In 1997, the Kentucky Towers was the largest residential building in Downtown Louisville.[3] In 2007 Downtown Louisville became Jefferson County's tenth Multiple Listing Service zone.

Housing units available downtown were expected to double between 2005 and 2010, from 1,800 to nearly 4,000, after increasing by only 900 units from 1985 to 2005.[4] This is both a result of new condominium construction and efforts to convert existing buildings into mixed usage, such as the $20 million redevelopment of the historic eight-story *Henry Clay* building at Third and Chestnut streets into a mix of residential, restaurant, retail and event space. The redevelopment also includes property that extends east to Fourth Street, which will become a public piazza, and the historic *Wright-Taylor* building, a two-story, 13500-square-foot (1250 m^2) structure that faces Fourth Street and is located behind the Henry Clay, and is now an upscale restaurant that occupies the entire Wright-Taylor building.

Liberty Greens will contain over 600 mixed income housing units

Future plans

Projects in the works include the conversion of the former Big Four railroad bridge into a pedestrian and bicycle only bridge, the construction of a wharf along the Riverwalk Trail, and the Ohio River Bridges Project, involving the reconstruction of Spaghetti Junction (the intersection of I-65, I-64 and I-71) along with the addition of a new bridge for northbound I-65 traffic.

On August 19, 2007, city leaders and the Cordish Company, developers of 4th Street Live!, announced *Center City*, a $442 million, multi-year plan to develop 23 acres (93000 m^2), bounded by Second, Third and Liberty streets and Muhammad Ali Boulevard, that will include new housing, restaurants, a cinema and a boutique hotel. An estimated 500000 square feet (46000 m^2) of floor space being created, including a 15-story structure. As the plan would require $130 million in local and state tax rebates for Cordish, it requires approval from the Louisville Metro Council and Kentucky General Assembly. There is no official start time for the project, as financing is still being secured by the Cordish Company.[5]

Also announced in 2007, the glass and steel $50 million shopping and office complex *Iron Quarter* was to be constructed within the Whiskey Row Historic District, but the project was delayed and eventually set aside when property owner Todd Blue made an agreement with the city of Louisville in January 2011 to demolish the seven original buildings. In May 2011, after all seven buildings had been landmarked, a new agreement was made with the city to save five of the seven buildings by donating one and selling the other four, with the remaining two to be demolished. Facades of all seven buildings are to be preserved.[6]

Attractions

Many attractions are located in Downtown Louisville.

- "Museum Row" in the West Main District
  - Frazier International History Museum
  - Kentucky Museum of Art and Craft
  - Louisville Glassworks
  - Louisville Science Center
  - Louisville Slugger Museum
  - Muhammad Ali Center
  - Museum Plaza (planned)
- East Market District, featuring a row of art galleries, prominently featured in the monthly First Friday Trolley Hop [7]
- Belle of Louisville
- Early Times Distillery
- Fort Nelson Park
- Fourth Street Live!
- The Kentucky Center
- Louisville Extreme Park
- Louisville Slugger Field (Home of the Louisville Bats)
- Waterfront Park
- Riverfront Plaza/Belvedere
- KFC Yum! Center (Home of the Louisville Cardinals)

Fourth Street Live!

Thunder Over Louisville

Louisville Slugger Bat factory and museum

Louisville Extreme Park

Muhammad Ali Center

Frazier International History Museum

Louisville Glassworks Museum and Artists Studio

Images

Modern high rises

The distances to each of Louisville's sister cities are represented on this lightpost downtown.

Many of Louisville's skyscrapers, from left: The Humana Building, National City Tower, LG&E Center (distant) and AEGON Center

AEGON Center, Kentucky's tallest building since 1993

The 800, Louisville's first modern high rise

PNC Tower

Waterfront Plaza I & II

Top Left: Waterfront Park Place (2004), Right: Preston Pointe (2004)

Jewish Hospital Medical Towers

The Marriott and Hyatt Regency Hotels in Dowtown

Construction on medical research building at Hancock and Liberty Streets

Construction on new hotel complex, Preston and Liberty Streets

Tallest buildings in Downtown Louisville

Rank	Building	Height	Floors	Year completed/projected	Status
1st	Museum Plaza	703 ft (214 m)	62	2012	Project Cancelled
2nd	AEGON Center (Capital Holding Center, Providian Center)	549 ft (167 m)	35	1993	Completed
3rd	National City Tower (First National Tower)	512 ft (156 m)	40	1972	Completed
4th	PNC Plaza (Citizens Fidelity Plaza)	420 ft (128 m)	30	1971	Completed
5th	Humana Building	417 ft (127 m)	27	1984	Completed
6th	Waterfront Park Place	364 ft (111 m)	23	2004	Completed
7th	Meidinger Tower (South Tower)	363 ft (110.6 m)	26	1982	Completed
8th	Brown & Williamson Tower (North Tower, Oxford Tower)	363 ft (110.6 m)	26	1982	Completed
9th	Waterfront Plaza I	340 ft (103.6 m)	25	1991	Completed
10th	Waterfront Plaza II	340 ft (103.6 m)	25	1993	Completed
11th	LG&E Center (One Corporate Plaza, LG&E Building)	328 ft (100 m)	23	1989	Completed
12th	Galt House (West Tower)	325 ft (99 m)	25	1972	Completed
13th	Galt House (East Tower)	322	20	1985	Completed
14th	BB&T Building (Louisville Trust Building, United Kentucky Building, Liberty Bank Building)	312 ft (95 m)	24	1972	Completed
15th	The 800 Apartments	290 ft (m)	29	1963	Completed
16th	Avenue Plaza Apartments Metro Housing Authority		18	1974	Completed
16th	Heyburn Building	250 ft (76 m)	17	1928	Completed
17th	J O Blanton House U.S. Housing and Urban Development		20	1972	Completed
18th	Hyatt Regency	246 ft (75 m)	18	1978	Completed
19th	Dosker Manor East and West Metro Housing Authority		18	1968	Completed
20th	Kentucky Home Life Building	235 ft (71.6 m)	19	1913	Completed
21st	Brown Hotel		16	1923	Completed
22nd	Kentucky Towers	202 ft (61.5 m)	18	1924	Completed
23rd	Starks Building	202 ft (61.5 m)	14	1913	Completed
24th	River View Place	201 ft (61 m)	19	1925	
25th	Marriott Louisville	200 ft (61 m)	17	2005	Completed

26th	ZirMed Gateway Towers		12	2009	Completed

See also

* Cityscape of Louisville, Kentucky
* Geography of Louisville, Kentucky
* List of tallest buildings in Louisville
* Ohio River Bridges Project

References

[1] Wiser, Stephen A.. *History of Louisville Architectural Firms* (http://www.aia-ckc.org/download.php?tn=mod_downloads_1&pk=fileid& id=138). . Retrieved 2010-04-28.

[2] *Louisville Guide* (2004), p. 136

[3] Nold, Chip; Bahr, Bob (1997). *Insiders' Guide to Louisville, Kentucky & Southern Indiana, 2nd* (http://books.google.com/ books?id=LEOE3_1PUgwC&pg=PA278&dq="kentucky+towers"&hl=en&ei=N9CWTLemGcP48Ab739iSDA&sa=X&oi=book_result& ct=result&resnum=2&ved=0CCgQ6AEwAQ#v=onepage&q="kentucky towers"&f=false). Globe Pequot. p. 278. ISBN 9781573800433. . Retrieved 2010-10-22.

[4] LouisvilleKy.gov - 2006 - Louisville's Downtown Alive with Development (http://www.louisvilleky.gov/DowntownDevelopment/News/ 2006/DowntownActivity.htm)

[5] Davis, Alex (2007-08-19). "More in store for downtown" (http://pqasb.pqarchiver.com/courier_journal/access/1724595761. html?FMT=ABS&FMTS=ABS:FT&type=current&date=Aug+19,+2007&author=&pub=Courier+-+Journal&edition=&startpage=A. 1&desc=More+in+store+for+downtown). The Courier-Journal. .

[6] LouisvilleKY.gov - Fischer Announces Plans to Preserve Historic Whiskey Row (http://www.louisvilleky.gov/Mayor/News/2011/ 5-9-11+whsikey+row+saved.htm)

[7] http://www.firstfridaytrolleyhop.com/

Further reading

* Yater, George H. (1987). *Two Hundred Years at the Falls of the Ohio: A History of Louisville and Jefferson County* (2nd edition ed.). Filson Club, Incorporated. ISBN 0-9601072-3-1.

External links

* Street Map of Downtown Louisville (http://www.ecentral.com/louisvillemaps/central_business_district.gif)
* Getting around Downtown Louisville (http://web.archive.org/web/20030506084827/http://www.nfb.org/ bm/bm03/bm0304/bm030405.htm)
* List of tallest buildings in Louisville (http://www.emporis.com/en/wm/ci/bu/?id=101641)
* Louisville Central Area website (http://web.archive.org/web/20080611005110/http://www.lca-inc.org/)
* Gallery of black-and-white photos of a nearly abandoned West Main Street in the 1970s (http://memory.loc. gov/cgi-bin/ampage?collId=pphhphoto&fileName=ky/ky0000/ky0005/photos/browse.db&action=browse& recNum=0&title2=Main Street, 600 &700 Block (Buildings), Louisville, Jefferson County, KY& displayType=1&itemLink=D?hh:5:./temp/~pp_I7zE::)
* Images of Central Business District / Downtown (Louisville, Ky.) in the University of Louisville Libraries Digital Collections (http://digital.library.louisville.edu/cdm4/results.php?CISORESTMP=results.php& CISOVIEWTMP=item_viewer.php&CISOMODE=grid& CISOGRID=thumbnail,A,1;title,A,1;descri,200,0;none,200,0;none,A,0;20;title,none,none,none,none& CISOBIB=title,A,1,N;subjec,A,0,N;descri,200,0,N;none,A,0,N;none,A,0,N;20;title,none,none,none,none& CISOTHUMB=20 (4x5);title,none,none,none,none&CISOTITLE=20;title,none,none,none,none& CISOHIERA=20;subjec,title,none,none,none&CISOSUPPRESS=0&CISOTYPE=link&CISOOP1=exact& CISOFIELD1=title&CISOBOX1=&CISOOP2=exact&CISOFIELD2=coveraa&CISOBOX2=Central+

Business+District+(Louisville,+Ky.)&CISOOP3=exact&CISOFIELD3=descri&CISOBOX3=&
CISOOP4=exact&CISOFIELD4=CISOSEARCHALL&CISOBOX4=&c=exact&CISOROOT=all)
- Louisville Downtown Management District - Interactive Downtown Map (http://www.ldmd.org/living.html)
- Louisville Downtown Management District recognizes 2010 development projects – Feb 2011 (http://www.
bizjournals.com/louisville/news/2011/02/24/louisville-downtown-management.html)

Kentucky

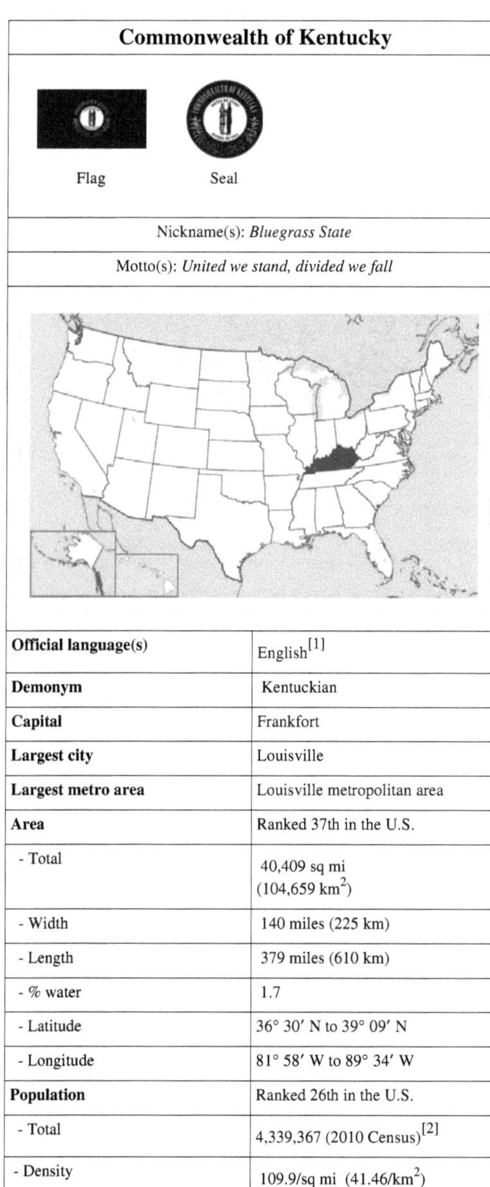

Commonwealth of Kentucky	
Flag	Seal
Nickname(s): *Bluegrass State*	
Motto(s): *United we stand, divided we fall*	
Official language(s)	English[1]
Demonym	Kentuckian
Capital	Frankfort
Largest city	Louisville
Largest metro area	Louisville metropolitan area
Area	Ranked 37th in the U.S.
- Total	40,409 sq mi (104,659 km^2)
- Width	140 miles (225 km)
- Length	379 miles (610 km)
- % water	1.7
- Latitude	36° 30′ N to 39° 09′ N
- Longitude	81° 58′ W to 89° 34′ W
Population	Ranked 26th in the U.S.
- Total	4,339,367 (2010 Census)[2]
- Density	109.9/sq mi (41.46/km^2) Ranked 24th in the U.S.
Elevation	

- Highest point	Black Mountain[3] 4,145 ft (1,263 m)
- Mean	755 ft (230 m)
- Lowest point	Kentucky Bend[3] 257 ft (78 m)
Admission to Union	June 1, 1792 (15th)
Governor	Steve Beshear (D)
Lieutenant Governor	Daniel Mongiardo (D)
Legislature	General Assembly
- Upper house	Senate
- Lower house	House of Representatives
U.S. Senators	Mitch McConnell (R) Rand Paul (R)
U.S. House delegation	4 Republicans, 2 Democrats (list)
Time zones	
- eastern half	Eastern: UTC-5/DST-4
- western half	Central: UTC-6/DST-5
Abbreviations	KY US-KY
Website	[Kentucky.gov Kentucky.gov]

The **Commonwealth of Kentucky** ◀ [i]/kɪnˈtʌki/ is a state located in the East Central United States of America. As classified by the United States Census Bureau, Kentucky is a Southern state, more specifically in the East South Central region. Kentucky is one of four U.S. states constituted as a commonwealth (the others being Virginia, Pennsylvania, and Massachusetts). Originally a part of Virginia, in 1792 it became the 15th state to join the Union. Kentucky is the 37th largest state in terms of total area, the 36th largest in land area, and ranks 26th in population.

Kentucky is known as the "Bluegrass State", a nickname based on the fact that bluegrass is present in many of the pastures throughout the state, based on the fertile soil. It made possible the breeding of high-quality livestock, especially thoroughbred racing horses. It is a land with diverse environments and abundant resources, including the world's longest cave system, Mammoth Cave National Park; the greatest length of navigable waterways and streams in the contiguous United States; and the two largest man-made lakes east of the Mississippi River. It is also home to the highest per capita number of deer and turkey in the United States, the largest free-ranging elk herd east of Montana, and the nation's most productive coalfield. Kentucky is also known for horse racing, bourbon distilleries, bluegrass music, automobile manufacturing, tobacco and college basketball.

Origin of name

It is generally accepted that the historic Native American tribes who hunted in what is now Kentucky referred to the region as *Catawba,* or some similar variant. The origin of Kentucky's modern name (variously spelled *Cane-tuck-ee, Cantucky, Kain-tuck-ee,* and *Kentuckee* before its modern spelling was accepted)[4] comes from an Iroquois word meaning "meadow lands", referring to the buffalo hunting grounds in Central Kentucky's savanna. Members of the *Haudenosaunee,* the Iroquois Confederacy, were historically based in New York and Pennsylvania. They penetrated to this area of the Ohio River Valley and drove other tribes out in order to control more hunting land. In addition to buffalo, they trapped beaver for the lucrative fur trade with the French and English, long before European-American settlement in this area.[5]

Narrow country roads bounded by stone and wood plank fences are a fixture in the Kentucky Bluegrass region.

Geography

Kentucky is considered to be situated in the Upland South. It is infrequently included in the Midwest.[6] [7] A significant portion of eastern Kentucky is part of Appalachia.

Kentucky borders seven states, from the Midwest and the Southeast. West Virginia lies to the east, Virginia to the southeast, Tennessee to the south, Missouri to the west, Illinois and Indiana to the northwest, and Ohio to the north and northeast. Only Missouri and Tennessee, both of which border eight states, touch more states.

Kentucky's northern border is formed by the Ohio River and its western border by the Mississippi River. The official state borders are

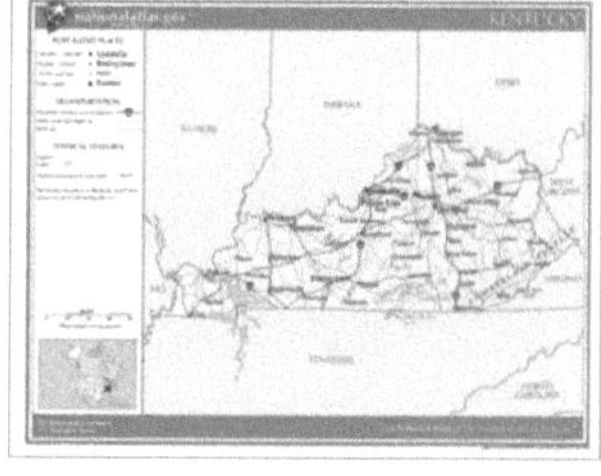

based on the courses of the rivers as they existed when Kentucky became a state in 1792. In several places, the rivers have changed courses away from the original borders. For instance, northbound travelers on US 41 from Henderson, after crossing the Ohio River, will be in Kentucky for about a half-mile (800 m) longer on the north side. Ellis Park, a thoroughbred racetrack, is located in this small piece of Kentucky. Waterworks Road is part of the only land border between Indiana and Kentucky.[8]

Kentucky is the only U.S. state to have a non-contiguous part existing as an exclave surrounded by other states. Fulton County, in the far west corner of the state, includes Kentucky Bend. This small part of Kentucky on the Mississippi River, bordered by Missouri and accessible via Tennessee, was created by the 1812 New Madrid Earthquake changing the course of the river.[9]

Regions

Kentucky can be divided into five primary regions: the Cumberland Plateau in the east, the north-central Bluegrass region, the south-central and western Pennyroyal Plateau, the Western Coal Fields and the far-west Jackson Purchase. The Bluegrass region is commonly divided into two regions, the Inner Bluegrass—the encircling 90 miles (145 km) around Lexington—and the Outer Bluegrass—the region that contains most of the Northern portion of the state, above the

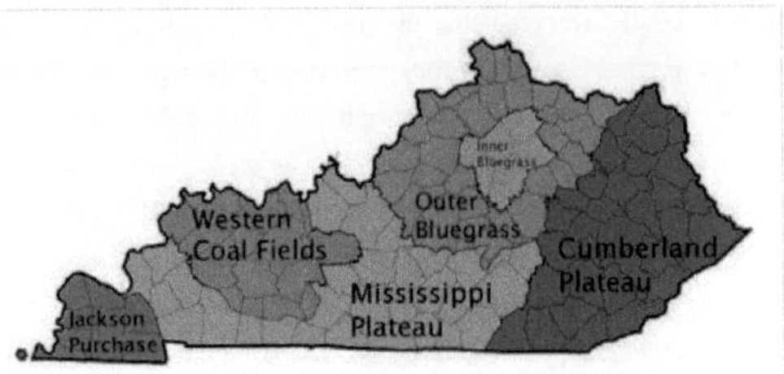

Kentucky's regions (click on image for color coding information.)

Knobs. Much of the outer Bluegrass is in the Eden Shale Hills area, made up of short, steep, and very narrow hills. This map is a rough depiction of the regions because it relies largely on county lines; as a result, the Inner Bluegrass appears larger than it is, and the Cumberland Plateau appears slightly smaller. The latter region is more commonly known in Kentucky as the East Kentucky Coal Field. Note the singular; these regions are not the sites of coal "fields" but one continuous field with many overlapping seams; the West Kentucky Coal Field is part of the Illinois Basin.

Kentucky's Inner Bluegrass region features hundreds of horse farms

The Jackson Purchase and western Pennyrile are home to several bald cypress/tupelo swamps

The East Kentucky Coal Field is known for its rugged terrain

Climate

Located within the southeastern interior portion of North America, Kentucky has a climate that can best be described as a humid subtropical climate (Koppen *Cfa*). Monthly average temperatures in Kentucky range from a summer daytime high of 87 °F (31 °C) to a winter low of 23 °F (−5 °C). The average precipitation is 46 inches (1200 mm) a year.[10] Kentucky experiences all four seasons, usually with striking variations in the severity of summer and winter from year to year.[11] Kentucky's highest recorded temperature was 114 °F (46 °C) at Greensburg on July 28, 1930 while the lowest recorded temperature was −34 °F (−37 °C) at Cynthiana on January 28, 1963.

Major weather events that have affected Kentucky include:

Event	Death Toll
Louisville Tornado of 1890	est. 76–120+
Ohio River flood of 1937	?
April 3, 1974 Tornado Outbreak	72
April 7, 1977 Flooding (Cumberland River toppled Pineville floodwall)	?
March 1, 1997 Flooding [12]	18
North American blizzard of 2003	?
2008 Super Tuesday tornado outbreak	Weather.com reported 17 deaths
September 2008 Windstorm	1
January 2009 ice storm	24+

Monthly Average High and Low Temperatures For Various Kentucky Cities												
City	Jan	Feb	Mar	Apr	May	Jun	Jul	Aug	Sep	Oct	Nov	Dec
Lexington	40/24	45/28	55/36	65/44	74/54	82/62	86/66	85/65	78/58	67/46	54/37	44/28
Louisville	41/25	47/28	57/37	67/46	75/56	83/65	87/70	86/68	79/61	68/48	56/39	45/30
Paducah	42/24	48/28	58/37	68/46	77/55	85/64	89/68	87/65	81/57	71/45	57/36	46/28
Pikeville	46/23	50/25	60/32	69/39	77/49	84/58	87/63	86/62	80/56	71/42	60/33	49/26
Ashland	42/19	47/21	57/29	68/37	77/47	84/56	88/61	87/59	80/52	69/40	57/31	46/23

Lakes and rivers

Kentucky's 90000 miles (km) of streams provides one of the most expansive and complex stream systems in the nation. Kentucky has both the largest artificial lake east of the Mississippi in water volume (Lake Cumberland) and surface area (Kentucky Lake). It is the only U.S. state to be bordered on three sides by rivers—the Mississippi River to the west, the Ohio River to the north, and the Big Sandy River and Tug Fork to the east.[13] Its major internal rivers include the Kentucky River, Tennessee River, Cumberland River, Green River and Licking River.

Though it has only three major natural lakes,[14] the state is home to many artificial lakes. Kentucky also has more navigable miles of water than any other state in the union, other than Alaska.[15]

Lake Cumberland is the largest artificial lake, in volume, east of the Mississippi River.

Natural environment and conservation

Once an industrial wasteland, Louisville's reclaimed waterfront now features thousands of trees and miles of walking trails

Kentucky has an expansive park system which includes one national park, two National Recreation areas, two National Historic Parks, two national forests, two National Wildlife Refuges, 45 state parks, 37696 acres (153 km^2) of state forest, and 82 Wildlife Management Areas.

Kentucky has been part of two of the most successful wildlife reintroduction projects in United States history. In the winter of 1997, the Kentucky Department of Fish and Wildlife Resources began to re-stock elk in the state's eastern counties, which had been extinct from the area for over 150 years. As of 2009, the herd had reached the project goal of 10,000 animals, making it the largest herd east of the Mississippi River.[16]

The state also stocked wild turkeys in the 1950s. Once extinct here, more wild turkeys thrive in Kentucky today than in any other eastern state. Hunters telechecked a record 29,006 birds taken during the 23-day season in Spring 2009. Only bearded turkeys were allowed to be hunted; the females are protected for breeding.[17]

In March 2011, Kentucky was rated last place amongst the states in the American State Litter Scorecard, presented at the American Society for Public Administration national conference.[18]

Significant natural attractions

- Cumberland Gap, chief passageway through the Appalachian Mountains in early American history.
- Cumberland Falls State Park, the only place in the Western Hemisphere where a "moon-bow" may be regularly seen, due to the spray of the falls.[19]
- Mammoth Cave National Park, featuring the world's longest known cave system.[20]
- Red River Gorge Geological Area, part of the Daniel Boone National Forest.
- Land Between the Lakes, a National Recreation Area managed by the United States Forest Service.
- Big South Fork National River and Recreation Area near Whitley City.

Red River Gorge is one of Kentucky's most visited places

- Black Mountain, state's highest point.[3] Runs along the border of Harlan and Letcher counties.
- Bad Branch Falls State Nature Preserve, 2639-acre (11 km^2) state nature preserve on southern slope of Pine Mountain in Letcher County. Includes one of the largest concentrations of rare and endangered species in the state,[21] as well as a 60-foot (18 m) waterfall and a Kentucky Wild River.
- Jefferson Memorial Forest, located in the southern fringes of Louisville in the Knobs region, the largest municipally run forest in the United States.[22]
- Lake Cumberland, 1255 miles (2020 km) of shoreline located in South Central Kentucky.
- Natural Bridge, located in Slade, Kentucky Powell County
- Breaks Interstate Park, located in southeastern Pike County, Kentucky and Southwestern Virginia. The Breaks is commonly known as the "Grand Canyon of the South."

History

Although inhabited by Native Americans from at least 1000 BC to about 1650 AD, when European and colonial explorers and settlers began entering Kentucky in greater number in the mid-18th century, there were no major Native American settlements in the region. The Shawnees from the north and Cherokees from the south sent parties into the area regularly for hunting. As more settlers entered the area, warfare broke out because the American Indians saw settlers' attempts to own land to be encroachment on their traditional hunting grounds.[23] The Southern Cherokee Nation of Kentucky still lives in Kentucky today as a state Recognized Tribe.

Abraham Lincoln Birthplace near Hodgenville

According to a 1790 U.S. government report, 1,500 Kentucky settlers had been killed in Indian raids since the end of the Revolutionary War.[24] In an attempt to end these raids, Clark led an expedition of 1,200 drafted men against Shawnee towns on the Wabash River in 1786, one of the first actions of the Northwest Indian War.[25]

After the American Revolution, the counties of Virginia beyond the Appalachian Mountains became known as Kentucky County.[26] Eventually, the residents of Kentucky County petitioned for a separation from Virginia. Ten constitutional conventions were held in the Constitution Square Courthouse in Danville

Both Abraham Lincoln and Jefferson Davis were born in Kentucky.

between 1784 and 1792. In 1790, Kentucky's delegates accepted Virginia's terms of separation, and a state constitution was drafted at the final convention in April 1792. On June 1, 1792, Kentucky became the fifteenth state to be admitted to the union. Isaac Shelby, a military veteran from Virginia, was elected the first Governor of the Commonwealth of Kentucky.[27]

Kentucky was a border state during the American Civil War.[28] Although frequently described as never having seceded, representatives from several counties met at Russellville calling themselves the "Convention of the People of Kentucky" and passed an Ordinance of Secession on November 20, 1861.[29] They established a Confederate government of Kentucky with its capital in Bowling Green.[30] Though Kentucky was represented by the central star on the Confederate battle flag,[31] the Russellville Convention did not represent the majority of residents. A year earlier, philosopher Karl Marx wrote to Friedrich Engels that the result of a vote deciding how Kentucky would be represented at a convention of the border states was "100,000 for the Union ticket, only a few thousand for secession."[32] Kentucky officially remained "neutral" throughout the war due to Union sympathies of many of the Commonwealth's citizens. Confederate Memorial Day is observed by some in Kentucky on Confederate President Jefferson Davis' birthday, June 3.[33]

Designed by the Washington Monument's architect Robert Mills in 1845, the U.S. Marine Hospital in Louisville is considered the best remaining antebellum hospital in the United States

The Black Patch Tobacco Wars, a vigilante action, occurred in the area in the early 20th century. As result of the tobacco industry monopoly, tobacco farmers in the area were forced to sell their tobacco at low prices. Many local farmers and activists united to refuse to sell tobacco to the tobacco industry. A vigilante wing, the "Night Riders", terrorized farmers who sold their tobacco at the low prices demanded by the tobacco corporations. They burned several tobacco warehouses, notably in Hopkinsville and Princeton. In the later period of their operation, they were known to physically assault farmers in the middle of the night who broke the boycott. The Governor declared martial law and deployed the Kentucky Militia to end the Black Patch Tobacco Wars.

On January 30, 1900, Governor William Goebel, flanked by two bodyguards and walking to the State Capitol in downtown Frankfort, was mortally wounded by an assassin. Goebel was contesting the election of 1899, which William S. Taylor was initially believed to have won. For several months, J. C. W. Beckham, Goebel's running mate, and Taylor fought over who was the legal governor, until the Supreme Court of the United States ruled in May in favor of Beckham. After fleeing to Indiana, Taylor was indicted as a co-conspirator in Goebel's assassination. Goebel is the only governor of a U.S. state to have been assassinated while in office.[34]

Law and government

Kentucky is one of four U.S. states to officially use the term commonwealth, which it inherited from Virginia. Kentucky is also one of only five states that elects its state officials in odd-numbered years (the others are Louisiana, Mississippi, New Jersey, and Virginia). Kentucky holds elections for these offices every 4 years in the years preceding Presidential election years. Thus, the last year when Kentucky elected a Governor was 2007; the next gubernatorial election will occur in 2011, with future gubernatorial elections to take place in 2015, 2019, 2023, etc.

Executive branch

The executive branch is headed by the governor who serves as both head of state and head of government. The lieutenant governor may or may not have executive authority depending on whether the person is a member of the Governor's cabinet. Under the current Kentucky Constitution, the lieutenant governor assumes the duties of the governor only if the governor is

The governor's mansion in Frankfort, Kentucky

incapacitated. (Prior to 1992, the lieutenant governor assumed power any time the governor was out of the state.) The governor and lieutenant governor usually run on a single ticket (also per a 1992 constitutional amendment), and are elected to four-year terms. Currently, the governor and lieutenant governor are Democrats Steve Beshear and Daniel Mongiardo.

Other elected constitutional offices include: the Secretary of State, Attorney General, Auditor of Public Accounts, State Treasurer and Commissioner of Agriculture. Currently, Democrat Elaine N. Walker serves as the Secretary of State. The commonwealth's chief prosecutor, law enforcement officer, and law officer is the attorney general. The current Kentucky attorney general is Democrat Jack Conway. The Auditor of Public Accounts is held by Democrat Crit Luallen. Democrat Todd Hollenbach is the current Treasurer. Former University of Kentucky basketball player Richie Farmer, a Republican, serves as the current Commissioner of Agriculture.

Legislative branch

Kentucky's legislative branch consists of a bicameral body known as the Kentucky General Assembly.

The Senate is considered the upper house. It has 38 members, and is led by the President of the Senate, currently Republican David L. Williams.

The House of Representatives has 100 members, and is led by the Speaker of the House, currently Democrat Greg Stumbo.

The Kentucky State Capitol building in Frankfort

Judicial branch

The judicial branch of Kentucky is called the Kentucky Court of Justice [35] and comprises courts of limited jurisdiction called District Courts; courts of general jurisdiction called Circuit Courts; specialty courts such as Drug Court, Family Court [36]; an intermediate appellate court, the Kentucky Court of Appeals; and a court of last resort, the Kentucky Supreme Court.

The Kentucky Court of Justice is headed by the Chief Justice of the Commonwealth.

Unlike federal judges, who are usually appointed, justices serving on Kentucky state courts are chosen by the state's populace in non-partisan elections.

Federal representation

Kentucky's two Senators are Senate Minority Leader Mitch McConnell and Rand Paul, both Republicans. The state is divided into six Congressional Districts, represented by Republicans Ed Whitfield (1st), Brett Guthrie (2nd), Geoff Davis (4th), and Hal Rogers (5th), and Democrats John Yarmuth (3rd) and Ben Chandler (6th).

Judicially, Kentucky is split into two Federal court districts: the Kentucky Eastern District and the Kentucky Western District. Appeals are heard in the Sixth Circuit Court of Appeals based in Cincinnati, Ohio.

A map showing Kentucky's six congressional districts

Law

Kentucky's body of laws, known as the Kentucky Revised Statutes (KRS), were enacted in 1942 to better organize and clarify the whole of Kentucky law.[37] The statutes are enforced by local police, sheriffs and deputy sheriffs, and constables and deputy constables. Unless they have completed a police academy elsewhere, these officers are required to complete training at the Kentucky Department of Criminal Justice Training Center on the campus of Eastern Kentucky University.[38] Additionally, in 1948, the Kentucky General Assembly established the Kentucky State Police, making it the 38th state to create a force whose jurisdiction extends throughout the given state.[39]

Kentucky is one of 36 states in the United States that sanctions the death penalty for certain crimes. Those convicted of capital crimes after March 31, 1998 are always executed by lethal injection; those convicted before this date may opt for the electric chair.[40] Only three people have been executed in Kentucky since the U.S. Supreme Court re-instituted the practice in 1976. The most notable execution in Kentucky, however, was that of Rainey Bethea on

August 14, 1936. Bethea was publicly hanged in Owensboro for the rape and murder of Lischia Edwards.[41] Irregularities with the execution led to this becoming the last public execution in the United States.[42]

Kentucky has been on the front lines of the debate over displaying the Ten Commandments on public property. In the 2005 case of *McCreary County v. ACLU of Kentucky*, the U.S. Supreme Court upheld the decision of the Sixth Circuit Court of Appeals that a display of the Ten Commandments in the Whitley City courthouse of McCreary County was unconstitutional.[43] Later that year, Judge Richard Fred Suhrheinrich, writing for the Sixth Circuit Court of Appeals in the case of *ACLU of Kentucky v. Mercer County*, wrote that a display including the Mayflower Compact, the Declaration of Independence, the Ten Commandments, the Magna Carta, *The Star-Spangled Banner*, and the national motto could be erected in the Mercer County courthouse.[44]

Kentucky has also been known to have unusually high political candidacy age laws, especially compared to surrounding states. The origin of this is unknown, but it has been suggested it has to do with the commonwealth tradition.

Politics

Further information: Political party strength in Kentucky

Presidential elections results[45]

Year	Republicans	Democrats
2008	**57.37%** *1,048,462*	41.15% *751,985*
2004	**59.55%** *1,069,439*	39.69% *712,733*
2000	**56.50%** *872,492*	41.37% *638,898*
1996	44.88% *623,283*	**45.84%** *636,614*
1992	41.34% *617,178*	**44.55%** *665,104*
1988	**55.52%** *734,281*	43.88% *580,368*
1984	**60.04%** *822,782*	39.37% *539,589*
1980	**49.07%** *635,274*	47.61% *616,417*
1976	45.57% *531,852*	**52.75%** *615,717*
1972	**63.37%** *676,446*	34.77% *371,159*
1968	**43.79%** *462,411*	37.65% *397,541*
1964	35.65% *372,977*	**64.01%** *669,659*
1960	**53.59%** *602,607*	46.41% *521,855*

Where politics are concerned, Kentucky historically has been very hard fought and leaned slightly toward the Democratic Party, although it was never included among the "Solid South". In 2006, 57.05% of the state's voters were officially registered as Democrats, 36.55% registered Republican, and 6.39% registered with some other political party.[46] Despite this, the state often supports Republican candidates for federal offices.

From 1964 through 2004, Kentucky voted for the eventual winner of the election for President of the United States. In the 2008 election, however, the state lost its bellwether status when John McCain, who won Kentucky, lost the national popular and electoral vote to Barack Obama (McCain carried Kentucky 57 to 41%). The Commonwealth supported the previous three Democratic candidates elected to the White House, all elected from Southern states: Lyndon B. Johnson (Texas) in 1964, Jimmy Carter (Georgia) in 1976, and Bill Clinton (Arkansas) in 1992 and 1996. In presidential elections, the state has become a Republican stronghold, supporting that party's presidential candidates by double-digit margins in 2000, 2004 and 2008.

Voter Registration and Party Enrollment as of June 26, 2010[47]			
Party		Number of Voters	Percentage
	Democratic	1,619,391	56.59%
	Republican	1,052,902	36.79%
	Other	189,499	6.62%
Total		2,861,792	100%

Demographics

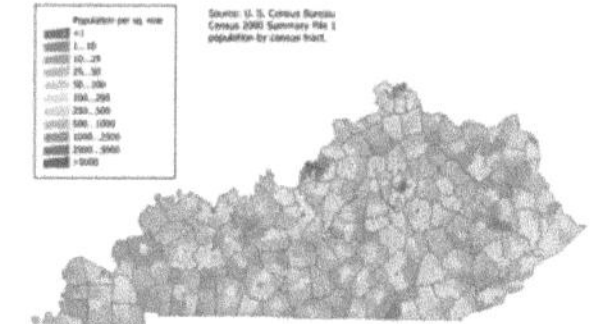

Kentucky Population Density Map.

Historical populations			
Census	Pop.		%±
1790	73677		—
1800	220955		199.9%
1810	406511		84.0%
1820	564317		38.8%
1830	687917		21.9%
1840	779828		13.4%
1850	982405		26.0%
1860	1155684		17.6%
1870	1321011		14.3%
1880	1648690		24.8%
1890	1858635		12.7%
1900	2147174		15.5%
1910	2289905		6.6%
1920	2416630		5.5%
1930	2614589		8.2%
1940	2845627		8.8%
1950	2944806		3.5%
1960	3038156		3.2%
1970	3218706		5.9%
1980	3660777		13.7%

1990	3685296		0.7%
2000	4041769		9.7%
2010	4339367		7.4%
Source: 1790-2000[48] 1910-2010[49]			

As of July 1, 2006, Kentucky has an estimated population of 4,206,074, which is an increase of 33,466, or 0.8%, from the prior year and an increase of 164,586, or 4.1%, since the year 2000. This includes a natural increase since the last census of 77,156 people (that is 287,222 births minus 210,066 deaths) and an increase due to net migration of 59,604 people into the state. Immigration from outside the United States resulted in a net increase of 27,435 people, and migration within the country produced a net increase of 32,169 people. As of 2004, Kentucky's population included about 95,000 foreign-born persons (2.3%). The population density of the state is 101.7 people per square mile.[50]

Kentucky's total population has grown during every decade since records began. However, during most decades of the 20th century there was also net out-migration from Kentucky. Since 1900, rural Kentucky counties have experienced a net loss of over 1 million people from migration, while urban areas have experienced a slight net gain.[51]

The center of population of Kentucky is located in Washington County, in the city of Willisburg.[52]

Race and ancestry

The largest ancestries in the commonwealth are: English (30.6%), German (12.7%), Irish (10.5%), and African American (7.8%).[53] [54] In the state's most urban counties of Jefferson, Oldham, Fayette, Boone, Kenton, and Campbell, German is the largest reported ancestry. Americans of Scots-Irish and English stock are present throughout the entire state, and many claim Irish ancestry because of the term "Scots-Irish", but most of the time in Kentucky this term is used for those with Scottish roots, rather than Irish.[55] Southeastern Kentucky was populated by a large group of Native Americans of mixed heritage, also known as Melungeons, in the early 19th century. Groups like the Ridgetop Shawnee were organizing the descendants of those early Native American settlers.

African Americans, who made up one-fourth of Kentucky's population prior to the Civil War, declined in number as many moved to the industrial North in the Great Migration. Today, the African American population of Jefferson County is 20%; 44.2% of Kentucky's African American population is in Jefferson County and 52% are in the Louisville Metro Area. Other areas with high concentrations, besides Christian and Fulton Counties, are the city of Paducah, the Bluegrass, and the city of Lexington. Many mining communities in far Southeastern Kentucky also have populations between five and 10 percent African American.

Demographics of Kentucky (csv) [56]

By race	White	Black	AIAN*	Asian	NHPI*
2000 (total population)	91.53%	7.76%	0.61%	0.92%	0.08%
2000 (Hispanic only)	1.35%	0.10%	0.04%	0.02%	0.01%
2005 (total population)	91.27%	7.98%	0.58%	1.10%	0.08%
2005 (Hispanic only)	1.80%	0.12%	0.04%	0.03%	0.01%
Growth 2000–05 (total population)	2.97%	6.16%	-2.21%	23.46%	9.78%
Growth 2000–05 (non-Hispanic only)	2.44%	5.94%	-3.28%	23.07%	7.98%
Growth 2000–05 (Hispanic only)	37.97%	22.34%	13.51%	38.48%	19.80%
* AIAN is American Indian or Alaskan Native; NHPI is Native Hawaiian or Pacific Islander					

Religion

In 2000, The Association of Religion Data Archives reported[57] that of Kentucky's 4,041,769 residents:

- 47% were not affiliated with any church
- 34% were members of Evangelical Protestant churches
  - Southern Baptist Convention (979,994 members, 24%)
  - Christian churches and churches of Christ (106,638 members, 3%)
  - Churches of Christ (58,602 members, 1%)
- 10% were Roman Catholics
- 9% belonged to mainline Protestant churches
  - United Methodist Church (208,720 members, 5%)
  - Christian Church (Disciples of Christ) (67,611 members, 2%)
- 0.05% were Orthodox Christians

Lexington Theological Seminary (then College of the Bible), 1904.

- 1% were affiliated with other theologies.

Today Kentucky is home to several seminaries. Southern Baptist Theological Seminary in Louisville is the principal seminary for the Southern Baptist Convention. Louisville is also the home of the Louisville Presbyterian Theological Seminary. Lexington has two seminaries, Lexington Theological Seminary, and the Baptist Seminary of Kentucky. Asbury Theological Seminary is located in nearby Wilmore. In addition to seminaries, there are several colleges affiliated with denominations. Transylvania in Lexington is affiliated with the Disciples of Christ. The University of Pikeville in Pikeville, Kentucky is affiliated with the Presbyterian Church. In Louisville, Bellarmine and Spalding are affiliated with the Roman Catholic Church. In Owensboro, Kentucky, Kentucky Wesleyan College is associated with the Methodist Church and Brescia University is associated with the Roman Catholic Church. Wilmore is home to Asbury University (a separate institution from the seminary), which is associated with the Christian College Consortium. The University of the Cumberlands, located in Williamsburg, Campbellsville University in Campbellsville, Georgetown College in Georgetown and Mid-Continent University in Mayfield all have connections with the Southern Baptist Convention. Louisville is also home to the headquarters of the Presbyterian Church (USA) and their printing press. Louisville is also home to a sizable Muslim[58] and Jewish population.

Economy

Early in its history Kentucky gained recognition for its excellent farming conditions. It was the site of the first commercial winery in the United States (started in present day Jessamine County in 1799) and due to the high calcium content of the soil in the Bluegrass region quickly became a major horse breeding (and later racing) area. Today Kentucky ranks 5th nationally in goat farming, 8th in beef cattle production,[59] and 14th in corn production.[60]

Today Kentucky's economy has expanded to importance in non agricultural terms as well, especially in auto manufacturing, energy fuel production, and medical facilities.As of 2010 24% of electricity produced in the USA depended on either enriched uranium rods coming from the Paducah Gaseous Diffusion Plant (the only domestic site of low grade uranium enrichment), or from the 107,336 tons of coal extracted from the state's two coal fields (which combined produce 4% percent of the electricity in the United States).[61] Kentucky ranks 4th among U.S. states in the number of automobiles and trucks assembled.[62] The Chevrolet Corvette, Cadillac XLR (2004–2009), Ford Explorer, Ford Super Duty trucks, Ford Excursion (2000–2005), Toyota Camry, Toyota Avalon, Toyota Solara, and Toyota Venza are assembled in Kentucky.

The total gross state product for 2010 was $163.3 billion, 28th in the nation.[63] Its per-capita personal income was US$28,513, 43rd in the nation.[64]

As of October 2010, the state's unemployment rate is 10%.[65]

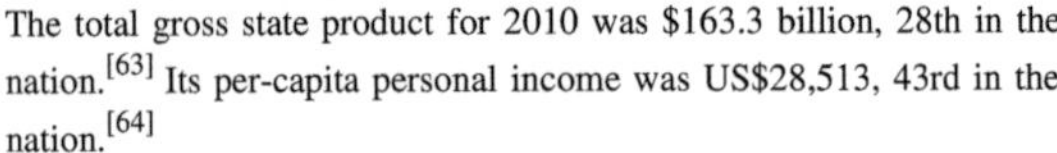
The best selling car in the United States, the Toyota Camry, is manufactured in Georgetown, Kentucky.

The best selling truck in the United States, the Ford F-Series, is manufactured in Louisville, Kentucky.

Taxation

There are six income tax brackets, ranging from 2% to 6% of personal income.[66] The sales tax rate in Kentucky is 6%.[67] Kentucky has a broadly based classified property tax system. All classes of property, unless exempted by the Constitution, are taxed by the state, although at widely varying rates.[68] Many of these classes are exempted from taxation by local government. Of the classes that are subject to local taxation, three have special rates set by the General Assembly, one by the Kentucky Supreme Court and the remaining classes are subject to the full local rate, which includes the tax rate set by the local taxing bodies plus all voted levies. Real property is assessed on 100% of the fair market value and property taxes are due by December 31. Once the primary source of state and local government revenue, property taxes now account for only about 6% of the Kentucky's annual General Fund revenues.[69]

Until January 1, 2006, Kentucky imposed a tax on intangible personal property held by a taxpayer on January 1 of each year. The Kentucky intangible tax was repealed under House Bill 272.[70] Intangible property consisted of any property or investment which represents evidence of value or the right to value. Some types of intangible property included: bonds, notes, retail repurchase agreements, accounts receivable, trusts, enforceable contracts sale of real estate (land contracts), money in hand, money in safe deposit boxes, annuities, interests in estates, loans to stockholders, and commercial paper.

"Unbridled Spirit"

Kentucky state welcome sign

To boost Kentucky's image, give it a consistent reach, and help Kentucky "stand out from the crowd", former Governor Ernie Fletcher launched a comprehensive branding campaign with the hope of making its $12 – $14 million advertising budget more effective. The "Unbridled Spirit" brand was the result of a $500,000 contract with New West, a Kentucky-based public relations advertising and marketing firm to develop a viable brand and tag line. The Fletcher administration aggressively marketed the brand in both the public and private sectors. The "Welcome to Kentucky" signs at border areas have Unbridled Spirit's symbol on them.

The previous campaign was neither a failure nor a success. Kentucky's "It's that friendly" slogan hoped to draw more people into the state based on the idea of southern hospitality. Though it was meant to embrace southern values, most Kentuckians rejected it as cheesy and ineffective. It was quickly seen that it was also not an image that encouraged tourism as much as initially hoped for. Therefore it was necessary to reconfigure a slogan to embrace Kentucky as a whole while also encouraging more people to visit the Bluegrass.[71]

Transportation

Main article: Transportation in Kentucky.

Roads

Kentucky is served by five major interstate highways (I-75, I-71, I-64, I-65, I-24), nine parkways, and three bypasses and spurs. The parkways were originally toll roads, but on November 22, 2006, Governor Ernie Fletcher ended the toll charges on the William H. Natcher Parkway and the Audubon Parkway, the last two parkways in Kentucky to charge tolls for access.[72] The related toll booths have been demolished.[73]

Ending the tolls some seven months ahead of schedule was generally agreed to have been a positive economic development for transportation in Kentucky. In June 2007, a law went into effect raising the speed limit on rural portions of Kentucky Interstates from 65 to 70 miles per hour (105 to 110 km/h).[74]

Greyhound provides bus service to most major towns in the state.

At 464 miles (747 km) long, Kentucky Route 80 is the longest route in Kentucky, pictured here west of Somerset.

The current state license plate design, introduced in 2005.

Rails

High Bridge over the Kentucky River was the tallest rail bridge in the world when it was completed in 1877.

Amtrak, the national passenger rail system, provides service to Ashland, South Portsmouth, Maysville and Fulton. The *Cardinal* (trains 50 and 51) is the line that offers Amtrak service to Ashland, South Shore, Maysville and South Portsmouth. The *City of New Orleans* (trains 58 and 59) serve Fulton. The Northern Kentucky area is served by the *Cardinal* at the Cincinnati Museum Center at Union Terminal. The Museum Center is just across the Ohio River in Cincinnati.

As of 2004, there were approximately 2640 miles (4250 km) of railways in Kentucky, with about 65% of those being operated by CSX Transportation. Coal was by far the most common cargo, accounting for 76% of cargo loaded and 61% of cargo delivered.[75]

Bardstown features a tourist attraction known as *My Old Kentucky Dinner Train*. Run along a 20-mile (30 km) stretch of rail purchased from CSX in 1987, guests are served a four-course meal as they make a two-and-a-half hour round-trip between Bardstown and Limestone Springs.[76] The Kentucky Railway Museum is located in nearby New Haven.[77]

Other areas in Kentucky are reclaiming old railways in rail trail projects. One such project is Louisville's Big Four Bridge. If completed, the Big Four Bridge rail trail will contain the second longest pedestrian-only bridge in the world.[78] The longest pedestrian-only bridge is also found in Kentucky—the Newport Southbank Bridge, popularly known as the "Purple People Bridge", connecting Newport to Cincinnati, Ohio.[79]

Air

Kentucky's primary airports include Louisville International Airport (Standiford Field), Cincinnati/Northern Kentucky International Airport (CVG), and Blue Grass Airport in Lexington. Louisville International Airport is home to UPS's Worldport, its international air-sorting hub.[80] There are also a number of regional airports scattered across the state.

On August 27, 2006, Kentucky's Blue Grass Airport in Lexington was the site of a crash that killed 47 passengers and 2 crew members aboard a Bombardier Canadair Regional Jet designated Comair Flight 191, or Delta Air Lines Flight 5191, sometimes mistakenly identified by the press as Comair Flight 5191.[81] The lone survivor was the flight's first officer, James Polehinke, who doctors determined to be brain damaged and unable to recall the crash at all.[82]

Water

Being bounded by two of the largest rivers in North America, water transportation has historically played a major role in Kentucky's economy. Most barge traffic on Kentucky waterways consists of coal that is shipped from both the Eastern and Western Coalfields, about half of which is used locally to power many power plants located directly off the Ohio River, with the rest being exported to other countries, most notably Japan.

Many of the largest ports in the United States are located in or adjacent to Kentucky, including:

A barge hauling coal in the Louisville and Portland Canal, the only manmade section of the Ohio River

- Huntington/Tri-State (includes Ashland, KY), largest inland port and 7th largest overall
- Cincinnati-Northern Kentucky, 5th largest inland port and 43rd overall
- Louisville-Southern Indiana, 7th largest inland port and 55th overall

As a state, Kentucky ranks 10th overall in port tonnage.[83] [84]

The only natural obstacle along the entire length of the Ohio River was the Falls of the Ohio, located just west of Downtown Louisville.

Subdivisions and settlements

Counties

Kentucky is subdivided into 120 counties, the largest being Pike County at 787.6 square miles (2040 km^2), and the most populous being Jefferson County (which coincides with the Louisville Metro governmental area) with 741,096 residents as of 2010.[85]

County government, under the Kentucky Constitution of 1891, is vested in the County Judge/Executive), (formerly called the County Judge) who serves as the executive head of the county, and a legislature called a Fiscal Court. Despite the unusual name, the Fiscal Court no longer has judicial functions.

Consolidated city-county governments

Kentucky's two most populous counties, Jefferson and Fayette, have their governments consolidated with the governments of their largest cities. *Louisville-Jefferson County Government* (Louisville Metro) and *Lexington-Fayette Urban County Government* (Lexington Metro) are unique in that their city councils and county Fiscal Court structures have been merged into a single entity with a single chief executive, the Metro Mayor and Urban County Mayor, respectively. Although the counties still exist as subdivisions of the state, in reference the names Louisville and Lexington are used to refer to the entire area coextensive with the former cities and counties. Somewhat incongruously, when entering Lexington-Fayette the highway signs read "Fayette County" while most signs leading into Louisville-Jefferson simply read "Welcome to Louisville Metro."

Cities and towns

Rank	City	2010 Pop	2000 Pop	Δ Current Pop
1	Louisville	566,503	551,299	566,503
2	Lexington	295,803	260,512	295,803
3	Bowling Green	58,067	49,296	58,067
4	Owensboro	57,265	54,067	57,265
5	Covington	40,640	43,370	40,640
6	Hopkinsville	31,577	30,089	31,577
7	Richmond	31,364	27,152	31,364
8	Florence	29,951	23,551	29,951
9	Georgetown	29,098	18,080	29,098
10	Henderson	28,757	27,373	28,757
11	Elizabethtown	28,531	22,542	28,531
12	Nicholasville	28,015	19,680	28,015
13	Jeffersontown	26,595	26,442	26,595
14	Frankfort	25,527	27,741	25,527
15	Paducah	25,024	26,442	25,024

The Louisville Metro government area has a 2010 population of 741,096. Under United States Census Bureau methodology, the population of Louisville was 566,503. The latter figure is the population of the so-called "balance"—the parts of Jefferson County that were either unincorporated or within the City of Louisville before the formation of the merged government in 2003. In 2010, the Louisville Combined Statistical Area (CSA) has a population of 1,451,564; including 1,061,031 in Kentucky, which is nearly one-fourth of the state's population. Since

2000, over one-third of the state's population growth has occurred in the Louisville CSA. In addition, the top 28 wealthiest places in Kentucky are in Jefferson County and seven of the 15 wealthiest counties in the state are located in the Louisville CSA.[86]

The second largest city is Lexington with a 2010 census population of 295,803 and its CSA, which includes the Frankfort and Richmond statistical areas, having a population of 687,173. The Northern Kentucky area (the seven Kentucky counties in the Cincinnati MSA) had a population of 425,483 in 2010. The metropolitan areas of Louisville, Lexington, and Northern Kentucky have a combined population of 2,173,687 as of 2010, which is 50.1% of the state's total population.

The two other fast growing urban areas in Kentucky are the Bowling Green area and the "Tri Cities Region" of southeastern Kentucky, comprising Somerset, London and Corbin.

Although only one town in the "Tri Cities", namely Somerset, currently has more than 10,000 people, the area has been experiencing heightened population and job growth since the 1990s. Growth has been especially rapid in Laurel County, which outgrew areas such as Scott and Jessamine counties around Lexington or Shelby and Nelson Counties around Louisville. London significantly grew in population in the 2000s, from 5,692 in 2000 to 7,993 in 2010. London also landed a Wal-Mart distribution center in 1997, bringing thousands of jobs to the community.

In northeast Kentucky, the greater Ashland area is an important transportation, manufacturing, and medical center. Iron and petroleum production, as well as the transport of coal by rail and barge, have been historical pillars of the region's economy. Due to a decline in the area's industrial base, Ashland has seen a sizable reduction in its population since 1990. The population of the area has since stabilized, however, with the medical service industry taking a greater role in the local economy. The Ashland area, including the counties of Boyd and Greenup, are part of the Huntington-Ashland, WV-KY-OH, Metropolitan Statistical Area (MSA). As of the 2000 census, the MSA had a population of 288,649. More than 21,000 of those people (as of 2010) reside within the city limits of Ashland.

The largest county in Kentucky by area is Pike, which contains Pikeville and suburb Coal Run Village . The county and surrounding area is the most populated region in the state that is not part of a Micropolitan Statistical Area or a Metropolitan Statistical Area containing nearly 200,000 people in five counties: Floyd County, Martin County, Letcher County, and neighboring Mingo County, West Virginia. Pike County contains slightly over 68,000 people.

Only three U.S. states have capitals with smaller populations than Kentucky's Frankfort (pop. 25,527), those being Augusta, Maine (pop. 18,560), Pierre, South Dakota (pop. 13,876), and Montpelier, Vermont (pop. 8,035).

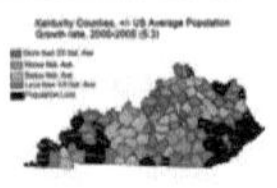

Population growth [87] is centered along and between interstates I-65 and I-75.

Louisville is the state's largest city with a metro population of 1.2 million.

Lexington is the state's second largest city with a metro population of around 500,000.

Although Covington, Kentucky only has a population of 40,000, the Kentucky side of the Cincinnati/Northern Kentucky metropolitan area has a population of over 400,000.

Newport's Aquarium and
waterfront

Education

Kentucky maintains eight public four-year universities. There are two general tiers: major research institutions (the University of Kentucky and the University of Louisville) and regional universities, which encompasses the remaining 6 schools. The regional schools have specific target counties that many of their programs are targeted towards (such as Forestry at Eastern Kentucky University or Cave Management at Western Kentucky University), however most of their curriculum varies little from any other public university. "UK" and "U of L" have the highest academic rankings and admissions standards although the regional schools aren't without their national recognized departments - examples being Western Kentucky University's nationally ranked Journalism Department or Morehead State offering one of the nation's only Space Science degrees. "UK" is the flagship and land grant of the system and has agriculture extension services in every county. The two research schools split duties related to the medical field, "UK" handles all medical outreach programs in the eastern half of the state while "U of L" does all medical outreach in the state's western half.

The University of Kentucky is Kentucky's
flagship university

The University of Louisville is Kentucky's urban
research university

The state's sixteen public two-year colleges have been governed by the Kentucky Community and Technical College System since the passage of the Postsecondary Education Improvement Act of 1997, commonly referred to as House Bill 1.[88] Prior to the passage of House Bill 1, most of these colleges were under the control of the University of Kentucky.

Transylvania University, located in Lexington, is the oldest university west of the Allegheny Mountains, founded in 1780. Transylvania is a liberal arts university, consistently ranked in the top tier in the country.

Berea College, located at the extreme southern edge of the Bluegrass below the Cumberland Plateau, was the first coeducational college in the South to admit both black and white students, doing so from its very establishment in 1855.[89] This policy was successfully challenged in the United States Supreme Court in the case of *Berea College v. Kentucky* in 1908.[90] This decision effectively segregated Berea until the landmark *Brown v. Board of Education* in 1954.

Kentucky has been the site of much educational reform over the past two decades. In 1989, the Kentucky Supreme Court ruled that the state's education system was unconstitutional.[91] The response of the General Assembly was passage of the Kentucky Education Reform Act (KERA) the following year. Years later, Kentucky has shown progress, but most agree that further reform is needed.[92]

Culture

Although Kentucky's culture is generally considered to be Southern, it is unique in that it is also influenced by the Midwest and Southern Appalachia in certain areas of the state. The state is known for bourbon and whiskey distilling, tobacco, horse racing, and college basketball. Kentucky is more similar to the Upland South in terms of ancestry which is predominantly American.[93] Nevertheless, during the 19th century, Kentucky did receive a substantial number of German immigrants, who settled mostly in the Midwest, along the Ohio River primarily in Louisville, Covington and Newport.[94] Only Maryland, Delaware and West Virginia have higher German ancestry percentages than Kentucky among Census-defined Southern states, although

Old Louisville is the largest Victorian Historic neighborhood in the United States.

Kentucky's percentage is closer to Arkansas and Virginia's than the previously named state's percentages. Scottish Americans, English Americans and Scotch-Irish Americans have heavily influenced Kentucky culture, and are present in every part of the state.[95] Kentucky was a slave state, and blacks once comprised over one-quarter of its population. However, it lacked the cotton plantation system and never had the same high percentage of African Americans as most other slave states. With less than 8% of its current population being black, Kentucky is rarely included in modern-day definitions of the Black Belt, despite a relatively significant rural African American population in the Central and Western areas of the state.[96] [97] [98] Kentucky adopted the Jim Crow system of racial segregation in most public spheres after the Civil War, but the state never disenfranchised African American citizens to the level of the Deep South states, and it peacefully integrated its schools after the 1954 *Brown v. Board of Education* verdict, later adopting the first state civil rights act in the South in 1966.[99]

The biggest day in horse racing, the Kentucky Derby, is preceded by the two-week Derby Festival[100] in Louisville. Louisville also plays host to the Kentucky State Fair,[101] the Kentucky Shakespeare Festival,[102] and Southern gospel's annual highlight, the National Quartet Convention.[103] Bowling Green, the state's third-largest city and home to the only assembly plant in the world that manufactures the Chevrolet Corvette,[104] opened the National Corvette Museum in 1994.[105] The fourth-largest city, Owensboro, gives credence to its nickname of "Barbecue Capital of the World" by hosting the annual International Bar-B-Q Festival.[106]

Old Louisville, the largest historic preservation district in the United States featuring Victorian architecture and the third largest overall,[107] hosts the St. James Court Art Show, the largest outdoor art show in the United States.[108] The neighborhood was also home to the Southern Exposition (1883–1887), which featured the first public display of Thomas Edison's light bulb,[109] and was the setting of Alice Hegan Rice's novel, *Mrs. Wiggs of the Cabbage Patch* and Fontaine Fox's comic strip, the "Toonerville Trolley.[110]

The more rural communities are not without traditions of their own, however. Hodgenville, the birthplace of Abraham Lincoln, hosts the annual Lincoln Days Celebration, and will also host the kick-off for the National Abraham Lincoln Bicentennial Celebration in February 2008. Bardstown celebrates its heritage as a major bourbon-producing region with the Kentucky Bourbon Festival.[111] (Legend holds that Baptist minister Elijah Craig invented bourbon with his black slave in Georgetown, but some dispute this claim.)[112] Glasgow mimics Glasgow, Scotland by hosting the Glasgow Highland Games, its own version of the Highland Games,[113] and Sturgis hosts "Little Sturgis", a mini version of Sturgis, South Dakota's annual Sturgis Motorcycle Rally.[114] The residents of tiny Benton even pay tribute to their favorite tuber, the sweet potato, by hosting Tater Day.[115] Residents of Clarkson in Grayson County celebrate their city's ties to the honey industry by celebrating the Clarkson Honeyfest.[116] The Clarkson Honeyfest is held the last Thursday, Friday and Saturday in September, and is the "Official State Honey Festival of Kentucky."

Music

The breadth of music in Kentucky is indeed wide, stretching from the Purchase to the eastern mountains.

Renfro Valley, Kentucky is home to Renfro Valley Entertainment Center and the Kentucky Music Hall of Fame and is known as "Kentucky's Country Music Capital," a designation given it by the Kentucky State Legislature in the late 1980s. The Renfro Valley Barn Dance was where Renfro Valley's musical heritage began, in 1939, and influential country music luminaries like Red Foley, Homer & Jethro, Lily May Ledford & the Original Coon Creek Girls, Martha Carson, and many others have performed as regular members of the shows there over the years. The Renfro Valley Gatherin' is today America's second oldest continually broadcast radio program of any kind. It is broadcast on local radio station WRVK and a syndicated network of nearly 200 other stations across the United States and Canada every week.

Contemporary Christian music star Steven Curtis Chapman is a Paducah native, and Rock and Roll Hall of Famers The Everly Brothers are closely connected with Muhlenberg County, where older brother Don was born. Kentucky was also home to Mildred and Patty Hill, the Louisville sisters credited with composing the tune to the ditty Happy Birthday to You in 1893; Loretta Lynn (Johnson County), and Billy Ray Cyrus (Flatwoods). However, its depth lies in its signature sound—Bluegrass music. Bill Monroe, "The Father of Bluegrass", was born in the small Ohio County town of Rosine, while Ricky Skaggs, Brian Littrell and Kevin Richardson of the Backstreet Boys, Keith Whitley, David "Stringbean" Akeman, Louis Marshall "Grandpa" Jones, Sonny and Bobby Osborne, and Sam Bush (who has been compared to Monroe) all hail from Kentucky. The International Bluegrass Music Museum is located in Owensboro,[117] while the annual Festival of the Bluegrass is held in Lexington.[118]

Kentucky is also home to famed jazz musician and pioneer, Lionel Hampton (although this has been disputed in recent years).[119] Blues legend W.C. Handy and R&B singer Wilson Pickett also spent considerable time in Kentucky. The R&B group Midnight Star and Hip-Hop group Nappy Roots were both formed in Kentucky, as were country acts The Kentucky Headhunters, Montgomery Gentry and Halfway to Hazard, The Judds, as well as Dove Award-winning Christian groups Audio Adrenaline (rock) and Bride (metal). Heavy Rock band Black Stone Cherry hails from rural Edmonton, Indie rock band My Morning Jacket with lead singer and guitarist Jim James also originated out of Louisville, on the local independent music Scene. Rock band Cage the Elephant is also from Bowling Green. The bluegrass groups Driftwood and Kentucky Rain, along with Nick Lachey of the pop band 98 Degrees are also from Kentucky.

In eastern Kentucky, old-time music carries on the tradition of ancient ballads and reels developed in historical Appalachia.

Cuisine

Kentucky's cuisine is generally similar to traditional southern cooking, although in some areas of the state it can blend elements of both the South and Midwest.[120] [121] One original Kentucky dish is called the Hot Brown, a dish normally layered in this order: toasted bread, turkey, bacon, tomatoes and topped with mornay sauce. It was developed at the Brown Hotel in Louisville.[122] The Pendennis Club in Louisville is the birthplace of the *Old Fashioned* cocktail. Also, western Kentucky is known for its own regional style of barbecue.

Harland Sanders originated Kentucky Fried Chicken at his service station in North Corbin, though the first franchised KFC was located in South Salt Lake, Utah.[123]

Sports

Kentucky is the home of several sports teams such as Minor League Baseball's Class A Lexington Legends and AAA Louisville Bats. They are also home to the Frontier Leagues Florence Freedom and several teams in the MCFL. The Lexington Horsemen and Louisville Fire of the af2 appear to be interested in making a move up to the "major league" Arena Football League. The northern part of the state lies across the Ohio River from Cincinnati, which is home to a National Football League team, the Bengals, and a Major League Baseball team, the Reds. It is not uncommon for fans to park in the city of Newport and use the Newport Southbank Pedestrian Bridge, locally known as the "Purple People Bridge," to walk to these games in Cincinnati. Also, Georgetown College in Georgetown is the location for the Bengals' summer training camp.[124]

Kentucky's Churchill Downs hosts the Kentucky Derby.

As in many states, especially those without major league professional sport teams, college athletics are very important. This is especially true of the state's three Division I Football Bowl Subdivision (FBS) programs, including the Kentucky Wildcats, the Western Kentucky University Hilltoppers, and the Louisville Cardinals. The Wildcats, Hilltoppers, and Cardinals are among the most tradition-rich college basketball teams in the United States, combining for nine championships and 22 NCAA Final Fours; and all three are on the lists of total all-time wins, wins per season, and average wins per season. The Kentucky Wildcats are particularly notable, leading all Division I programs in all time wins, win percentage, NCAA tournament appearances, and being second only to UCLA in NCAA championships. Louisville has also stepped onto the football scene in recent years, with eight straight bowl games, including the 2007 Orange Bowl. Western Kentucky, the 2002 national champion in Division I-AA football (now Football Championship Subdivision (FCS), completed its transition to Division I FBS football in 2009.

Ohio Valley Wrestling in Louisville was the primary location for training and rehab for WWE professional wrestlers from 2000 until February 2008, when WWE ended its relationship with OVW and moved all of its contracted talent to Florida Championship Wrestling.

The NASCAR Sprint Cup Series will also have a race at the Kentucky Speedway in Sparta, Kentucky, an hour away from Louisville. The race will be called the Quaker State 400. The NASCAR Nationwide Series and the Camping World Truck Series also race there.

State symbols

Insignia	Symbol	Binomial nomenclature	Year Adopted[125]
Official State Bird	Cardinal	*Cardinalis cardinalis*	1926
Official State Butterfly	Viceroy Butterfly	*Limenitis archippus*	1990
Official State Dance	Clogging		2001
Official State Beverage	Milk		2005
Official State Fish	Kentucky Spotted Bass	*Micropterus punctulatus*	2005
Official State Fossil	Brachiopod	undetermined	1986
Official State Flower	Goldenrod	*Soldiago gigantea*	1926
Official State Fruit	Blackberry	*Rubus allegheniensis*	2004

Official State Gemstone	Freshwater Pearl		1986
State Grass	Kentucky Bluegrass	*Poa pratensis*	Traditional
Official State Motto	*"United we stand, divided we fall"*		1942
Official State Latin Motto	*"Deo gratiam habeamus"* ("Let us be grateful to God")		2002
Official State Horse	Thoroughbred	*Equus caballus*	1996
Official State Mineral	Coal		1998
Official State Outdoor Musical	"The Stephen Foster Story" (now called "Stephen Foster - The Musical")		2002
Official State Instrument	Appalachian Dulcimer		2001
State Nickname	"The Bluegrass State"		Traditional
Official State Rock	Kentucky Agate		2000
Official State Slogan	"Kentucky: Unbridled Spirit"		2004[126]
Official State Soil	Crider Soil Series		1990
Official State Tree	Tulip Poplar	*Liriodendron tulipifera*	1994
Official Wild Animal Game Species	Gray Squirrel	*Sciurus carolinensis*	1968
Official State Song	"My Old Kentucky Home" (revised version)		1986
Official State Silverware Pattern	*Old Kentucky Blue Grass: The Georgetown Pattern*		1996
Official State Music	Bluegrass music		2007[127]
Official State Automobile	Chevrolet Corvette		2010

Official state places and events

- State arboretum: Bernheim Arboretum and Research Forest
- State botanical garden: The Arboretum: State Botanical Garden of Kentucky
- State Science Center: Louisville Science Center
- State center for celebration of African American heritage: Kentucky Center for African American Heritage
- State honey festival: Clarkson Honeyfest[128]
- State amphitheater: Iroquois Amphitheater (Louisville)

- State tug-o-war championship: The Fordsville Tug-of-War Championship
- Covered Bridge Capital of Kentucky: Fleming County
- Official Covered Bridge of Kentucky: Switzer Covered Bridge (Franklin County)
- Official steam locomotive of Kentucky: "Old 152" (located in the Kentucky Railway Museum in New Haven)
- Official pipe band: Louisville Pipe Band
- State bourbon festival: Kentucky Bourbon Festival, Inc., in Bardstown, Kentucky

Unless otherwise specified, all state symbol information is taken from Kentucky State Symbols [129].

Gallery

The world famous Louisville Slugger baseball bat is made in Kentucky. It holds the Guinness World Record for the largest bat.

Kentucky's 2001 commemorative quarter.

Thunder Over Louisville is the largest annual fireworks show in the world.

Kentucky's horse farms are world renowned.

The Daniel Boone National Forest.

The Ohio River forms the northern border of Kentucky.

Many Kentucky cities have historic areas near downtown, such as this example in Bowling Green.

US Highway 23 cuts through the rugged Cumberland Plateau near Pikeville.

See also

- List of National Register of Historic Places in Kentucky
- List of people from Kentucky
- Southern Cherokee Nation of Kentucky
- U.S. state

References

[1] "Kentucky State Symbols" (http://kdla.ky.gov/resources/KYSymbols.htm). Kentucky Department for Libraries and Archives. . Retrieved 2006-11-29.

[2] "2010 Census: Resident Population Data" (http://2010.census.gov/2010census/data/apportionment-pop-text.php). 2010. . Retrieved 2010-12-22.

[3] "Science In Your Backyard: Kentucky" (http://www.usgs.gov/state/state.asp?State=KY)). United States Geological Survey. . Retrieved 2006-11-29.

[4] "State Symbols". Encyclopedia of Kentucky. New York, New York: Somerset Publishers. 1987. ISBN 0403099811.

[5] "Native American Tribes of Kentucky" (http://www.native-languages.org/kentucky.htm)

[6] ed. in chief Frederick C. Mish (2003). *Merriam-Webster's Collegiate Dictionary* (http://books.google.com/?id=O78rzaI2XmUC& pg=RA1-PA1562) (11th ed.). Merriam–Webster. p. 1562. ISBN 9780877798095. .

[7] The North American Midwest: A Regional Geography. New York, New York: Wiley Publishers. 1955. ISBN 0901411931.

[8] "Map of [1494-1557] Waterworks Rd Evansville, IN" (http://www.mapquest.com/maps/map.adp?formtype=address&addtohistory=& address=[1494-1557] Waterworks Rd&city=Evansville&state=IN&zipcode=47713&country=US& location=sx5PfJLyNGOdfPC4XlmsmD4sYz8/cM/9UzxNAshApmBL3N63w0vZEKUJ7ZFErueQQVVT7jNm9im/ rMyLwvsq1tX3B0QnxqQNsp3LVvDC22VDK3WLmnQ83dOStm4oo36rfS7/gXA9L8+8CqYgeWZpmK5YKDtojM0V&ambiguity=1). . Retrieved 2009-01-01.

[9] "Life on the Mississippi" (http://www.ket.org/kentuckylife/800s/kylife804.html). Kentucky Educational Television. 2002-01-28. . Retrieved 2006-11-29.

[10] "The Geography of Kentucky - Climate" (http://www.netstate.com/states/geography/ky_geography.htm). NetState.com. 2006-06-15. . Retrieved 2006-11-29.

[11] "Geographical Configuration". Encyclopedia of Kentucky. New York, New York: Somerset Publishers. 1987. ISBN 0403099811.

[12] http://www.enquirer.com/flood/floodphotos.html

[13] Kleber, John E., ed (1992). "Rivers". The Kentucky Encyclopedia. Associate editors: Thomas D. Clark, Lowell H. Harrison, and James C. Klotter. Lexington, Kentucky: The University Press of Kentucky. ISBN 0813117720.

[14] Kleber, John E., ed (1992). "Lakes". The Kentucky Encyclopedia. Associate editors: Thomas D. Clark, Lowell H. Harrison, and James C. Klotter. Lexington, Kentucky: The University Press of Kentucky. ISBN 0813117720.

[15] "Corbin, Kentucky: A Fisherman's Paradise" (http://www.corbinkentucky.us/fishing.htm). Corbin, Kentucky Economic Development. . Retrieved 2006-11-29.

[16] "Elk Restoration Update and Hunting Information" (http://web.archive.org/web/20060926064155/http://fw.ky.gov/elkinfo. asp?lid=1653&NavPath=C117C147C301C547). Kentucky Department of Fish and Wildlife Resources. Archived from the original (http:// fw.ky.gov/elkinfo.asp?lid=1653&NavPath=C117C147C301C547) on 2006-09-26. . Retrieved 2006-12-09.

[17] "Hunters Take Record Number of Spring Turkey" (http://fw.ky.gov/newsrelease.asp?nid=542). *Kentucky Department of Fish and Wildlife Service.* .

[18] S. Spacek, "2011 American State Litter Scorecard: New Rankings for an Increasingly Environmentally Concerned Populace".

[19] "Cumberland Falls State Resort Park" (http://web.archive.org/web/20061005015537/http://parks.ky.gov/resortparks/cf/). Kentucky Department of Parks. 2005-10-19. Archived from the original (http://parks.ky.gov/findparks/resortparks/cf/) on 2006-10-05. . Retrieved 2006-11-29.

[20] "Mammoth Cave National Park" (http://www.nps.gov/maca/). National Park Service. 2006-10-12. . Retrieved 2006-11-29.

[21] "Bad Branch State Nature Preserve" (http://web.archive.org/web/20061024022402/http://www.naturepreserves.ky.gov/stewardship/ badbranch.htm). Kentucky State Nature Preserves Commission. Archived from the original (http://www.naturepreserves.ky.gov/ stewardship/badbranch.htm) on October 24, 2006. . Retrieved 2006-11-29.

[22] "Jefferson Memorial Forest" (http://www.memorialforest.com/). . Retrieved 2006-11-29.

[23] "The Presence" (http://www.merceronline.com/Native/native01.htm). *History of Native Americans in Central Kentucky.* Mercer County Online. . Retrieved 2006-11-29.

[24] James, James Alton (1928). *The Life of George Rogers Clark.* Chicago: University of Chicago Press. ISBN 0404035493.

[25] Harrison, Lowell H (1976; Reprinted 2001). *George Rogers Clark and the War in the West.* Lexington: University Press of Kentucky. ISBN 0-8131-9014-2.

[26] "About Kentucky" (http://search.ezilon.com/about-kentucky.html). Ezilon Search. . Retrieved 2006-11-29.

[27] "Constitution Square State Historic Site" (http://web.archive.org/web/20071011230008/http://danville-ky.com/attractions2. php?category=History+and+Museums). Danville-Boyle County Convention and Visitors Bureau. Archived from the original (http://www. danville-ky.com/attractions2.php?category=History and Museums) on 2007-10-11. . Retrieved 2006-11-29.

[28] "Border States in the Civil War" (http://www.civilwarhome.com/borderstates.htm). CivilWarHome.com. 2002-02-15. . Retrieved 2006-11-29.

[29] "Ordinances of Secession" (http://historicaltextarchive.com/sections.php?op=viewarticle&artid=170). Historical Text Archive. . Retrieved 2006-11-29.

[30] "Civil War Sites - Bowling Green, KY" (http://www.trailsrus.com/monuments/reg3/bowling_green.html). WMTH Corporation. . Retrieved 2006-11-29.

[31] Irby, Jr., Richard E.. "A Concise History of the Flags of the Confederate States of America and the Sovereign State of Georgia" (http:// ngeorgia.com/history/flagsofga.html). *About North Georgia.* Golden Ink. . Retrieved 2006-11-29.

[32] Marx, Karl (1861-07-05). "Marx To Engels In Manchester" (http://marxists.org/archive/marx/works/1861/letters/61_07_05.htm). Marxists Internet Archive. . Retrieved 2006-11-29.

[33] "KRS 2.110 Public Holidays" (http://www.lrc.ky.gov/KRS/002-00/110.PDF) (PDF). Kentucky General Assembly. . Retrieved 2006-11-29.

[34] "The Old State Capitol" (http://history.ky.gov/sub.php?pageid=23§ionid=8). Kentucky Historical Society. . Retrieved 2007-11-09.

[35] http://courts.ky.gov

[36] http://courts.ky.gov/stateprograms/

[37] "Reviser of Statutes Office - History and Functions" (http://www.lrc.ky.gov/statrev/revoff.htm). Kentucky Legislative Research Commission. . Retrieved 2006-12-27.

[38] "History of the DOCJT" (http://docjt.jus.state.ky.us/history.html). Kentucky Department of Criminal Justice. . Retrieved 2006-12-27.

[39] "History of the Kentucky State Police" (http://www.kentuckystatepolice.org/history.htm). Kentucky State Police. . Retrieved 2006-12-27.

[40] "Authorized Methods of Execution by State" (http://www.deathpenaltyinfo.org/methods-execution#state). Death Penalty Information Center. . Retrieved 2006-12-28.

[41] Long, Paul A (2001-06-11). "'The Last Public Execution in America'" (http://web.archive.org/web/20060117233210/http://www. kypost.com/2001/jun/11/bethea061101.html). *The Kentucky Post* (E. W. Scripps Company). Archived from the original (http://www.

kypost.com/2001/jun/11/bethea061101.html) on 2006-01-17. . Retrieved 2006-12-27.

[42] Montagne, Renee (2001-05-01). "The Last Public Execution in America" (http://www.npr.org/programs/morning/features/2001/apr/010430.execution.html). NPR. . Retrieved 2006-12-27.

[43] *"McCreary County v. ACLU of Kentucky"* (http://straylight.law.cornell.edu/supct/html/03-1693.ZS.html). Cornell University Law School. . Retrieved 2006-12-27.

[44] "Text of decision in *ACLU of Kentucky v. Mercer County*" (http://www.ca6.uscourts.gov/opinions.pdf/05a0477p-06.pdf) (PDF). . Retrieved 2006-12-27.

[45] Leip, David. "Presidential General Election Results Comparison - Kentucky" (http://uselectionatlas.org/RESULTS/compare.php?year=2008&fips=21&f=1&off=0&elect=0&type=state). US Election Atlas. . Retrieved December 31, 2009.

[46] "2006 General Election Registration Figures Set" (http://web.archive.org/web/20071212043450/http://kentucky.gov/Newsroom/sos/article61.htm). Kentucky Secretary of State. 2006-10-19. Archived from the original (http://kentucky.gov/Newsroom/sos/article61.htm) on 2007-12-12. . Retrieved 2006-11-30.

[47] "Voter Registration Statistics" (http://www.elect.ky.gov/stats/regstat.htm) (PDF). Kentucky State Board of Elections. . Retrieved 2010-07-07.

[48] http://ukcc.uky.edu/census/21.txt

[49] http://2010.census.gov/2010census/data/apportionment-pop-text.php

[50] John W. Wright, ed (2007). *The New York Times 2008 Almanac.* p. 178.

[51] Price, Michael. "Migration in Kentucky: Will the Circle Be Unbroken?" (http://www.kltprc.net/books/exploring/Chpt_3.htm). *Exploring the Frontier of the Future: How Kentucky Will Live, Learn and Work.* University of Louisville. pp. 5–10. . Retrieved 30 April 2007.

[52] "Population and Population Centers by State: 2000" (http://www.census.gov/geo/www/cenpop/statecenters.txt) (TXT). U.S. Census Bureau. . Retrieved 2006-12-27.

[53] http://www.epodunk.com/cgi-bin/popInfo.php?locIndex=18

[54] http://www.epodunk.com/cgi-bin/genInfo.php?locIndex=18

[55] Census 2000 Map - Top U.S. Ancestries by County

[56] http://www.census.gov/popest/states/asrh/tables/SC-EST2005-03-21.csv

[57] "State Membership Report" (http://www.thearda.com/mapsReports/reports/state/21_2000.asp). The Association of Religion Data Archives. 2000. . Retrieved 2006-12-27.

[58] Muslims in Louisville (http://www.irfi.org/articles/articles_101_150/muslims_in_louisville.htm)

[59] "2007 Rankings of States and Counties" (http://www.bamabeef.org/NewStateandCountyrankings05.htm). bamabeef.org. . Retrieved 1 M a y 2007.

[60] "Corn Production Detective" (http://web.archive.org/web/20070605025350/http://www.econedlink.org/lessons/EM453/docs/em453_Corn_Production_Det_Answers.pdf) (PDF). National Council on Economic Education. Archived from the original (http://www.econedlink.org/lessons/EM453/docs/em453_Corn_Production_Det_Answers.pdf) on 2007-06-05. . Retrieved 2007-05-03.

[61] Utah Geological Survey - U.S. Coal Production by State, 1994-2009 (http://geology.utah.gov/emp/energydata/statistics/coal2.0/pdf/T2.7.pdf)

[62] url=http://www.tradeandindustrydev.com/issues/article.asp?ID=66

[63] "GDP by State" (http://greyhill.com/gdp-by-state). . Retrieved 7 September 2011.

[64] Kentucky Cabinet for Economic Development - Kentucky Economy (http://www.thinkkentucky.com/kyedc/pdfs/kyecotrd.pdf)

[65] Bls.gov (http://www.bls.gov/lau/); Local Area Unemployment Statistics

[66] "Kentucky Income Tax Rates" (http://swz.salary.com/salarywizard/layouthtmls/swzl_statetaxrate_KY.html). salary. com. . Retrieved May 1, 2007.

[67] "Sales & Use Tax" (http://revenue.ky.gov/business/salesanduse.htm). Kentucky Department of Revenue. . Retrieved May 1, 2007.

[68] "Property Tax" (http://web.archive.org/web/20070403180310/http://revenue.ky.gov/business/proptax.htm). Kentucky Department of Revenue. Archived from the original (http://revenue.ky.gov/business/proptax.htm) on April 3, 2007. . Retrieved May 1, 2007.

[69] "State Taxes - Kentucky - Overview" (http://www.bankrate.com/yho/itax/edit/state/profiles/state_tax_Ky.asp). bankrate.com. . Retrieved 2007-05-01.

[70] "Text of the House Bill 272" (http://www.lrc.ky.gov/record/05rs/HB272.htm). State of Kentucky. . Retrieved August 10, 2007.

[71] "Unbridled Spirit→Information" (http://kentucky.gov/Pages/unbridledspirit.aspx). State of Kentucky. . Retrieved 2007-05-01.

[72] Stinnett, Chuck. "Fletcher:Tolls to end November 22" (http://web.archive.org/web/20061008233115/http://www.kctcs.net/todaysnews/index.cfm?tn_date=2006-09-28#6693). Archived from the original (http://www.kctcs.net/todaysnews/index.cfm?tn_date=2006-09-28#6693) on 2006-10-08. . Retrieved 2007-05-01.

[73] Stinnett, Chuck (2006-11-22). "Onlookers Cheer Booth Destruction at Ceremony" (http://www.courierpress.com/news/2006/nov/22/onlookers-cheer-booth-destruction-at-ceremony/). Courier Press. . Retrieved August 10, 2007.

[74] Steitzer, Stephanie (2007-06-26). "Many new laws go on books today" (http://www.courier-journal.com/apps/pbcs.dll/article?AID=2007706260437). Courier-Journal. .

[75] "Railroad Service in Kentucky" (http://www.aar.org/PubCommon/Documents/AboutTheIndustry/RRState_KY.pdf) (PDF). Association of American Railroads. . Retrieved 2007-05-01. Also, Norfolk Southern's main north-south line runs through central and southern Kentucky, starting in Cincinnati. Formerly the CNO&TP subsidiary of Southern Railway, it is NS's most profitable line.

[76] Knight, Andy. "On the Right Track - Kentucky Dinner Train serves up railroad nostalgia" (http://www.cincinnati.com/visitorsguide/stories/071100_dinnertrain.html). Cincinnati.com. . Retrieved 2007-05-01.

[77] "Kentucky Railway Museum" (http://www.kyrail.org/). . Retrieved 2007-05-01.

[78] Shafer, Sheldon (2007-03-05). "Bridges money may be shifted". Courier-Journal.

[79] Crowley, Patrick (April 23, 2003). "Meet the Purple People Bridge" (http://www.enquirer.com/editions/2003/04/20/loc_purplebridge20.html). Cincinnati Enquirer. . Retrieved 2007-05-01.

[80] "Fast Facts" (http://www.flylouisville.com/About-the-Airport/About-the-Airport.aspx). Louisville International Airport. . Retrieved 2007-09-11.

[81] Crash Kills 49 (http://web.archive.org/web/20061105013158/http://www.kentucky.com/mld/kentucky/news/special_packages/crash/15378422.htm)

[82] "Comair Crash Survivor Leaves Hospital" (http://www.cbsnews.com/stories/2006/10/03/national/main2059120.shtml). CBS. 2006-10-03. . Retrieved 2007-05-01.

[83] Top 20 Inland U.S. Ports for 2003 (http://www.iwr.usace.army.mil/ndc/wcsc/pdf/inlandport03f.pdf)

[84] CY 2001 Tonnage for Selected U.S. Ports by Port Tons (http://www.iwr.usace.army.mil/ndc/wcsc/portton01.htm)

[85] Kentucky Counties (http://www.uky.edu/KentuckyAtlas/kentucky-counties.html), University of Kentucky

[86] "Kentucky State Data Center" (http://ksdc.louisville.edu/). Ksdc.louisville.edu. . Retrieved 2009-08-04.

[87] http://www.census.gov/popest/counties/tables/CO-EST2005-02-21.csv

[88] "Postsecondary Education Improvement Act of 1997" (http://www.lrc.ky.gov/recarch/97ss/HB1/bill.doc). State of Kentucky. . Retrieved 2007-05-01.

[89] "Berea College:Learning, Labor, and Service" (http://www.diversityweb.org/digest/vol10no1/mendel.cfm). Diversity Web. . Retrieved 2007-05-01.

[90] Berea College v. Kentucky (http://www.brownat50.org/brownCases/PreBrownCases/BereavKty1908.html)

[91] "A Guide to the Kentucky Education Reform Act of 1990" (http://eric.ed.gov/ERICWebPortal/Home.portal?_nfpb=true&_pageLabel=RecordDetails&ERICExtSearch_SearchValue_0=ED327352&ERICExtSearch_SearchType_0=eric_accno&objectId=0900000b8004b71c). Education Resources Information Center. . Retrieved 2007-05-01.[Abstract of A Guide to the Kentucky Education Reform Act of 1990 - provided by Education Resources Information Center (ERIC)]

[92] Roeder, Phillip. "Education Reform and Equitable Excellence: The Kentucky Experiment" (http://www.kltprc.net/foresight/Chpt_37.htm). . Retrieved 2007-05-01.

[93] Brittingham, Angela & de la Cruz, G. Patricia (June 2004). "Ancestry 2000: Census 2000 Brief" (http://www.census.gov/prod/2004pubs/c2kbr-35.pdf) (PDF). United States Census Bureau. . Retrieved 28 June 2007.

[94] "Kentucky's German Americans in the Civil War" (http://kygermanscw.yolasite.com/the-story.php). Kygermanscw.yolasite.com. . Retrieved 2010-07-02.

[95] "2000 Census: Percent Reporting Any German Ancestry" (http://www.thearda.com/mapsReports/maps/map.asp?state=101&variable=494). . Retrieved 2007-07-20.

[96] Beale, Calvin (21 July 2004). "High Poverty in the Rural U.S. and South: Progress and Persistence in the 1990s" (http://web.archive.org/web/20070626232430/http://srdc.msstate.edu/poverty/ppts/cromartie.ppt) (PowerPoint). Archived from the original (http://srdc.msstate.edu/poverty/ppts/cromartie.ppt) on June 26, 2007. . Retrieved 28 June 2007.

[97] Womack, Veronica L. (23 July 2004). "The American Black Belt Region: A Forgotten Place" (http://web.archive.org/web/20070626232429/http://srdc.msstate.edu/poverty/ppts/womack.ppt) (PowerPoint). Archived from the original (http://srdc.msstate.edu/poverty/ppts/womack.ppt) on 26 June 2007. . Retrieved 28 June 2007.

[98] Unknown. "Identifying the "Black Belt" of Cash-Crop Production" (http://www.bowdoin.edu/~prael/branch/ex1/m4-black-belt.jpg) (JPEG Image). Bowdoin College. . Retrieved 28 June 2007.

[99] "Civil Rights and Women's Rights" (http://encarta.msn.com/encyclopedia_761554924_12/Kentucky.html). *Civil Rights and Women's Rights*. . Retrieved 2007-07-20.

[100] "Derby Festival Home Page" (http://www.derbyfestival.org/). . Retrieved 2011-05-13.

[101] "Kentucky State Fair" (http://www.kystatefair.org/). . Retrieved 2006-12-25.

[102] "Kentucky Shakespeare Festival Home Page" (http://www.kyshakes.org/). . Retrieved 2006-12-25.

[103] "National Quartet Convention Home Page" (http://www.natqc.com/). . Retrieved 2006-12-25.

[104] "National Corvette Museum press release" (http://web.archive.org/web/20071227111243/http://bg.ky.net/Corvette/newera.htm). Archived from the original (http://bg.ky.net/Corvette/newera.htm) on 2007-12-27. . Retrieved 2006-12-25.

[105] "National Corvette Museum Home Page" (http://www.corvettemuseum.com/). . Retrieved 2006-12-25.

[106] "Home Page of the International Barbecue Festival" (http://www.bbqfest.com/). . Retrieved 2006-12-25.

[107] "Stately Mansions Grace Old Louisville" (http://www.ajc.com/travel/content/travel/southeast/ky_stories/0305/09lvgetaway.html). Atlanta Journal Constitution. . Retrieved 2006-12-25.

[108] "St. James Court Art Show Home Page" (http://www.stjamescourtartshow.com/). . Retrieved 2006-12-25.

[109] "The Heart Line" (http://chfs.ky.gov/NR/rdonlyres/6AD56B4B-7551-4E34-AE5B-E067472C503E/0/October_2004.pdf) (PDF). Kentucky Commission on Community Volunteerism and Service. . Retrieved 2006-12-25.

[110] "Old Louisville and Literature" (http://www.oldlouisville.com/literature/). . Retrieved 2006-12-25.

[111] "Kentucky Bourbon Festival Home Page" (http://www.kybourbonfestival.com/). . Retrieved 2006-12-25.

[112] "How Bourbon Whiskey *Really* Got Its Famous Name" (http://www.straightbourbon.com/articles/ccname.html). . Retrieved 2006-12-25.

[113] "Glasgow, Kentucky Highland Games Home Page" (http://www.glasgowhighlandgames.com/). . Retrieved 2006-12-25.

[114] "Little Sturgis Rally Home Page" (http://www.littlesturgisrally.net/). . Retrieved 2006-12-25.

[115] "Tater Day Festival A Local Legacy" (http://www.americaslibrary.gov/cgi-bin/page.cgi/es/ky/tater_1). . Retrieved 2006-12-25.

[116] "Clarkson Honeyfest home page" (http://www.honeyfest.com/). . Retrieved 2007-05-12.

[117] "International Bluegrass Music Museum" (http://www.bluegrass-museum.org/). . Retrieved 2006-11-30.

[118] "Festival of the Bluegrass Home Page" (http://www.festivalofthebluegrass.com/). . Retrieved 2006-11-30.

[119] Voce, Steve (2002-09-02). "Obituary: Lionel Hampton" (http://findarticles.com/p/articles/mi_qn4158/is_20020902/ai_n12639955/pg_5). The Independent. . Retrieved 2007-06-03.

[120] "Southern Recipes - Southern Food and Recipes" (http://southernfood.about.com/od/southernregionalfood/Southern_Recipes_and_Regional_Specialties.htm). Southernfood.about.com. 2009-06-17. . Retrieved 2009-08-04.

[121] "International Institute of Culinary Arts" (http://web.archive.org/web/20080106091226/www.iicaculinary.com/iica-curr.htm#ac303). Archived from the original (http://www.iicaculinary.com/iica-ye2-sem1.htm#ac303) on 2008-01-06. .

[122] "Hot Brown Recipe" (http://web.archive.org/web/20070823065032/http://www.brownhotel.com/dining/hot-brown.html). Brown Hotel. Archived from the original (http://www.brownhotel.com/dining/hot-brown.html) on August 23, 2007. . Retrieved 2006-12-18.

[123] Jenifer K. Nii (2004). "Colonel's landmark KFC is mashed". Deseret Morning News. http://www.deseretnews.com/article/1,5143,595057690,00.html.Retrieved on October 28, 2007.

[124] "About the camp" (http://www.bengalscamp.com). BengalsCamp.com (http://www.bengalscamp.com/). . Retrieved 2006-12-18.

[125] "Kentucky's State Symbols" (http://kdla.ky.gov/resources/KYSymbols.htm). Kentucky Department of Libraries and Archives. . Retrieved 2006-12-18.

[126] "Unbridled Spirit Information" (http://www.kentucky.gov/unbridledspirit/info.htm). Kentucky.gov (http://kentucky.gov/). 2006-11-20. . Retrieved 2006-12-18.

[127] "HB71: An act designating bluegrass music as the official state music of Kentucky" (http://www.lrc.ky.gov/RECORD/07RS/HB71/bill.doc) (DOC). Legislative Research Commission. . Retrieved 2007-06-26.

[128] "KRS 2.099 - State Honey Festival" (http://www.lrc.ky.gov/KRS/002-00/099.PDF) (PDF). Kentucky General Assembly. . Retrieved 2006-12-18.

[129] http://kdla.ky.gov/resources/KYSymbols.htm

Bibliography

Politics

- Miller, Penny M. *Kentucky Politics & Government: Do We Stand United?* (1994) (http://www.questia.com/PM.qst?a=o&d=3096856)
- Jewell, Malcolm E. and Everett W. Cunningham, *Kentucky Politics* (1968)

History

Surveys and reference

- Bodley, Temple and Samuel M. Wilson. *History of Kentucky* 4 vols. (1928).
- Caudill, Harry M., *Night Comes to the Cumberlands* (1963). ISBN 0-316-13212-8
- Channing, Steven. *Kentucky: A Bicentennial History* (1977).
- Clark, Thomas Dionysius. *A History of Kentucky* (many editions, 1937–1992).
- Collins, Lewis. *History of Kentucky* (1880).
- Harrison, Lowell H. and James C. Klotter. *A New History of Kentucky* (1997).
- Kleber, John E. et al. *The Kentucky Encyclopedia* (1992), standard reference history.
- Klotter, James C. *Our Kentucky: A Study of the Bluegrass State* (2000), high school text
- Lucas, Marion Brunson and Wright, George C. *A History of Blacks in Kentucky* 2 vols. (1992).
- Notable Kentucky African Americans http://www.uky.edu/Subject/aakyall.html
- Share, Allen J. *Cities in the Commonwealth: Two Centuries of Urban Life in Kentucky* (1982).
- Wallis, Frederick A. and Hambleton Tapp. *A Sesqui-Centennial History of Kentucky* 4 vols. (1945).
- Ward, William S., *A Literary History of Kentucky* (1988) (ISBN 0-87049-578-X).

- WPA, *Kentucky: A Guide to the Bluegrass State* (1939) (http://www.questia.com/PM.qst?a=o&d=99066764), classic guide.
- Yater, George H. (1987). *Two Hundred Years at the Fall of the Ohio: A History of Louisville and Jefferson County* (2nd ed.). Filson Club, Incorporated. ISBN 0-9601072-3-1.

Specialized scholarly studies

- Bakeless, John. *Daniel Boone, Master of the Wilderness* (1989) (http://www.questia.com/PM.qst?a=o&d=105132587)
- Blakey, George T. *Hard Times and New Deal in Kentucky, 1929–1939* (1986)
- Coulter, E. Merton. *The Civil War and Readjustment in Kentucky* (1926)
- Davis, Alice. "Heroes: Kentucky's Artists from Statehood to the New Millennium" (2004)
- Ellis, William E. *The Kentucky River* (2000).
- Faragher, John Mack. *Daniel Boone* (1993)
- Fenton, John H. *Politics in the Border States: A Study of the Patterns of Political Organization, and Political Change, Common to the Border States: Maryland, West Virginia, Kentucky, and Missouri* (1957) (http://www.questia.com/PM.qst?a=o&d=6471305)
- Ireland, Robert M. *The County in Kentucky History* (1976)
- Klotter, James C.; Lowell Harrison, James Ramage, Charles Roland, Richard Taylor, Bryan S. Bush, Tom Fugate, Dixie Hibbs, Lisa Matthews, Robert C. Moody, Marshall Myers, Stuart Sanders and Stephen McBride (2005). Jerlene Rose. ed. *Kentucky's Civil War 1861–1865*. Back Home In Kentucky Inc. ISBN 0-9769231-1-4.
- Klotter, James C. *Kentucky: Portrait in Paradox, 1900–1950* (1992)
- Pearce, John Ed. *Divide and Dissent: Kentucky Politics, 1930–1963* (1987)
- Remini, Robert V. *Henry Clay: Statesman for the Union* (1991).
- Sonne, Niels Henry. *Liberal Kentucky, 1780–1828* (1939) (http://www.questia.com/PM.qst?a=o&d=100662336)
- Tapp, Hambleton and James C Klotter. *Kentucky Decades of Discord, 1865–1900* (1977)
- Townsend, William H. *Lincoln and the Bluegrass: Slavery and Civil War in Kentucky* (1955) (http://www.questia.com/PM.qst?a=o&d=61648056)
- Waldrep, Christopher *Night Riders: Defending Community in the Black Patch, 1890–1915* (1993) (http://www.questia.com/PM.qst?a=o&d=14459929) tobacco wars

External links

- Kentucky.gov: My New Kentucky Home (http://kentucky.gov/)
- Kentucky State and County Government Websites (http://www.n2genealogy.com/kentucky.html#government)
- Kentucky State Databases (http://wikis.ala.org/godort/index.php/Kentucky) - Annotated list of searchable databases produced by Kentucky state agencies and compiled by the Government Documents Roundtable of the American Library Association.
- Kentucky (http://www.dmoz.org/Regional/North_America/United_States/Kentucky/) at the Open Directory Project
- Kentucky Department of Tourism (http://www.kentuckytourism.com/)
- Kentucky travel guide from Wikitravel
- The Kentucky Highlands Project (http://kentuckyhighlands.com/)
- USGS real-time, geographic, and other scientific resources of Kentucky (http://www.usgs.gov/state/state.asp?State=KY)
- Energy & Environmental Data for Kentucky (http://tonto.eia.doe.gov/state/state_energy_profiles.cfm?sid=KY)

- Kentucky State Facts (http://www.ers.usda.gov/StateFacts/KY.htm)
- Kentucky: Unbridled Spirit (http://www.kentuckyunbridledspirit.com/)
- Kentucky Virtual Library (http://www.kyvl.org/)
- "Science In Your Backyard: Kentucky" (http://www.usgs.gov/state/state.asp?State=KY) U.S. Department of the Interior I U.S. Geological Survey, July 3, 2006, retrieved November 4, 2006
- U.S. Census Bureau Kentucky QuickFacts (http://quickfacts.census.gov/qfd/states/21000.html)
- OpenStreetMap has geographic data related to Kentucky (http://www.openstreetmap.org/browse/relation/161655)
- Kentucky (http://ballotpedia.org/wiki/index.php/Kentucky) at Ballotpedia

gag:Kentucky mrj:Кентукки

Louisville and Nashville Railroad

Louisville and Nashville Railroad	
Reporting mark	LN
Locale	Southern United States
Dates of operation	1850–1982
Successor	Seaboard System
Track gauge	4 ft 8 $\frac{1}{2}$ in (1435 mm) (standard gauge)
Headquarters	Louisville, Kentucky

The **Louisville and Nashville Railroad** (reporting mark **LN**) was a Class I railroad that operated freight and passenger services in the southeast United States.

Chartered by the state of Kentucky in 1850, the **L&N**, as it was generally known, grew into one of the great success stories of American business. Operating under one name continuously for 132 years, it survived civil war and economic depression and several waves of social and technological change. As one of the premier Southern railroads, the L&N extended its reach far beyond its namesake cities, ultimately building a network of nearly 7000 miles (11000 km) of track.

Early history and Civil War

Its first line extended barely south of Louisville, Kentucky, and in fact it took until 1859 to span the 180-odd miles to its second namesake city of Nashville. There were about 250 miles (400 km) of track in the system by the outbreak of the Civil War, and its strategic location, spanning the Union/Confederate lines, made it of great interest to both governments.

During the Civil War, different parts of the network were pressed into service by both armies at various times, and considerable damage from wear, battle, and sabotage occurred. However, the company benefited from being based in the Union state of Kentucky, and the fact that Nashville fell to Union forces within the first year of the war and remained in their hands for its duration. It profited from Northern haulage contracts for troops and supplies, paid in sound Federal "greenbacks," as opposed to the rapidly-depreciating Confederate dollars. After the war, it found that its Southern competitors were devastated to the point of collapse, and the general economic depression meant that labor and materials to repair its roads could be had fairly cheaply.

Buoyed by these fortunate circumstances, the firm began an expansion that never really stopped. Within thirty years the network reached from Ohio and Missouri to Louisiana and Florida. By 1884, the firm had such importance that it was included in the Dow Jones Transportation Average, the first American stock market index. It was so active a customer of the Rogers Locomotive and Machine Works, the country's second largest locomotive maker, that in 1879 the firm presented L&N with a free locomotive as a thank-you bonus.

Coal and capital in the Gilded Age

Union Station in Louisville, Kentucky

Railroads were much interested in coal, of course, as all locomotives were steam-powered, and wood-burning models had been found to be unsatisfactory. The L&N shrewdly guaranteed not only its own fuel sources but a steady revenue stream by pushing its lines into the difficult but coal-rich terrain of eastern Kentucky, and also well into northern Alabama. There the small town of Birmingham had recently been founded amidst undeveloped deposits of coal, iron ore and limestone, the basic ingredients of steel production. The arrival of L&N transport and investment capital helped create a great industrial city, and the South's first postwar urban success story. In the first half of the 20th century, the railroad's ready access to very high-grade coal eventually enabled it to boast the nation's longest non-stop run, nearly 500 miles (800 km) from Louisville to Montgomery, Alabama without refueling.

In the Gilded Age of the late 19th century, there were no such things as anti-trust or fair-competition laws and very little in the way of financial regulation. Business was a keen and mean affair, and the L&N proved a most formidable competitor. It could, and did, simply freeze out upstarts like the Tennessee Central Railway Company from critical infrastructure like urban stations. Where that wasn't possible, as with the Nashville, Chattanooga and St. Louis Railway (which was older than the L&N), it simply used its financial muscle—in 1880 it acquired a controlling interest in its chief competitor. A public outcry resulted from this, however, sufficient to convince the L&N directors that there were limits even to their power. They discreetly continued the NC&StL as a separate subsidiary, but now working in complement to, instead of in competition with, the L&N.

Somewhat ironically, in 1902 financial speculations by the financier J.P. Morgan delivered control of the L&N Railroad to its rival Atlantic Coast Line Railroad. Curiously, however, that company did not make any attempt to control L&N operations, and for many decades there were no consequences of this change.

The Twentieth Century

The World Wars brought heavy demand to the L&N. Its widespread and robust network coped well with the demands of war transport and production, and the resulting profits harked back to the boost it had received from the Civil War. In the postwar period, the line shifted gradually to diesel power, and the new streamlined engines pulled some of the most elegant passenger trains of the last great age of passenger rail, such as the *Dixie Flyer*, the *Humming Bird*, and the *Pan-American*.

Though well past its hundredth anniversary, the line was still growing. In 1957, the Nashville, Chattanooga & St. Louis was finally fully merged. In the 1960s, acquisitions in Illinois allowed a long-sought entry into

Louisville Terminus at Union Station with 11-story L&N Building on the left.

the premier railroad nexus of Chicago, and some of the battered remains of the old rival, the Tennessee Central, were sold to the L&N Railroad as well.

In 1971, the Seaboard Coast Line Railroad, the successor to the Atlantic Coast Line Railroad, purchased the remainder of the L&N shares it did not already own, and the company became a subsidiary. During this period, in common with other lines, the L&N was cutting back passenger service. Amtrak, the government-formed passenger railway service, took over the few remaining L&N passenger trains in 1971. In 1979, amid great lamentations in the press, it ceased passenger service to its namesake cities when Amtrak discontinued *The Floridian*, which had connected Louisville with Nashville, and thence to Florida via Birmingham.

By 1982, the railroad industry was consolidating quickly, and the Seaboard Coast Line absorbed the Louisville & Nashville Railroad entirely. The merged company became known as "SCL/L&N", "Family Lines", and was depicted as such on the railroad's rolling stock. During the next few years, several smaller acquisitions resulted in the creation of the Seaboard System Railroad. Yet more consolidation was ahead, and in 1986, the Seaboard System was merged into the C&O/B&O combined system known as the "Chessie System". The combined company became CSX Transportation (CSX), which now owns and operates all of the former Louisville and Nashville railroad lines.

Few industries have as large and devoted a body of historians and fans as railroading does, and the long and colorful saga of the Louisville & Nashville has generated much interest. A number of historical groups and publications devoted to the line exist, and L&N equipment is well represented in the popular model railroading hobby.

Passenger Operations

While the *Humming Bird* and *Pan-American* were two of the L&N's most popular passenger trains that ran entirely on its own lines, from Cincinnati to New Orleans, the railroad also hosted other named trains, including:

- *Azalean* (New York - New Orleans in conjunction with the Southern Railway and the West Point Route)
- *Crescent* (New York - New Orleans in conjunction with the Southern Railway and the West Point Route)
- *Dixie Flagler* (Chicago - Miami in conjunction with the Atlantic Coast Line Railroad and the Florida East Coast Railway)
- *Dixieland* (Chicago - Miami in conjunction with the Atlantic Coast Line Railroad and the Florida East Coast Railway)
- *Flamingo* (Cincinnati - Jacksonville in conjunction with the Southern Railway)
- *Georgian* (Originally St. Louis - Atlanta, later changed to Chicago-Atlanta)
- *Gulf Wind* (New Orleans - Jacksonville in conjunction with the Seaboard Air Line Railroad)
- *Piedmont Limited* (New York - New Orleans in conjunction with the Southern Railway and the West Point Route)
- *Southland* (Detroit - Florida in conjunction with the Southern Railway)
- *South Wind* (Chicago - Miami in conjunction with the Atlantic Coast Line Railroad and the Florida East Coast Railway)

References

- Owen, Thomas McAdory (1921). *History of Alabama and Dictionary of Alabama Biography, Volume 1* [1]. Chicago, IL: The S. J. Clarke Publishing Company.
- Herr, Kincaid A. (2000; reprint of 1964 edition). *The Louisville and Nashville Railroad 1850–1963*. University Press of Kentucky. ISBN 0813121841.
- Klein, Maury (2002). *History of the Louisville & Nashville Railroad*. University Press of Kentucky. ISBN 0813122635.

See also

- Family Lines System
- List of Louisville and Nashville Railroad precursors
- *Louisville & Nashville Railroad Co. v. Mottley* -- famous U.S. Supreme Court case involving the railroad
- Seaboard System

External links

- L&N in Georgia [2]
- L&N in Alabama [3]
- L&N Railroad Historical Society [4]
- L&N model photos [5]
- A 1933 article on the *Louisville and Nashville Railroad* [6]
- L&N Historic Railpark Museum [7]
- Finding Aid, Louisville and Nashville Railroad Company records [8], University of Louisville Archives and Records Center

References

[1] http://books.google.com/books?id=ZEkUAAAAYAAJ

[2] http://www.railga.com/ln.html

[3] http://web.archive.org/web/20071007200454/http://www.oldalabamarails.org/history3.html

[4] http://www.lnrr.org/

[5] http://appalachianrailroadmodeling.com/lnmodels.html

[6] http://www.nzetc.org/tm/scholarly/tei-Gov07_08Rail-t1-body-d16.html

[7] http://www.historicrailpark.com/

[8] http://kdl.kyvl.org/cgi/f/findaid/
    findaid-idx?c=klgead;id=navbarbrowselink;cginame=findaid-idx;cc=klgead;view=reslist;subview=standard;didno=klgarulrar052

Pilaster

A **pilaster** is a slightly-projecting column built into or applied to the face of a wall. Most commonly flattened or rectangular in form, pilasters can also take a half-round form or the shape of any type of column, including tortile.

In discussing Leon Battista Alberti's use of pilasters, which Alberti reintroduced into wall-architecture, Rudolf Wittkower wrote, "The pilaster is the logical transformation of the column for the decoration of a wall. It may be defined as a flattened column which has lost its three-dimensional and tactile value."[1]

A pilaster appears with a capital[2] and entablature, also in "low-relief" or flattened against the wall.

The pilaster is an architectural element in classical architecture used to give the appearance of a supporting column and to articulate an extent of wall, with only an ornamental function. In contrast, an engaged column or buttress can support the structure of a wall and roof above.

Paired Corinthian pilasters on the county courthouse in Springfield, Ohio.

Pilasters often appear on the sides of a door frame or window opening on the facade of a building, and are sometimes paired with columns or pillars set directly in front of them at some distance away from the wall, which support a roof structure above, such as a portico. These vertical elements can also be used to support a recessed archivolt around a doorway. The pilaster can be replaced by ornamental brackets supporting the entablature or a balcony over a doorway.

When a pilaster appears at the corner intersection of two walls it is known as a **canton**.[3]

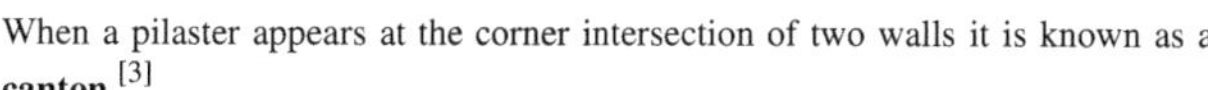

As with a column, a pilaster can have a plain or fluted surface to its profile and can be represented in the mode of any architectural style. During the Renaissance and Baroque architects a range of pilaster forms. [4] In the giant order pilasters appear as two-storeys tall, linking floors in a single unit.

The fashion of using this element from Ancient Greek and Roman architecture was adopted in the Italian Renaissance, gained wide popularity with Greek Revival architecture, and continues to be seen in some modern architecture.

See also

- Archivolt
- Buttress
- Post and lintel
- Engaged column
- Classical architecture
- List of classical architecture terms

Notes

[1] Wittkower, *Architectural Principles in the Age of Humanism* (1962) 1965:36.

[2] A useful phrase to identify a section of pilaster without a capital, with only its fluting to identify its relation to a column, is "pilaster strip".

[3] Ching, Francis D. K. (1995). A Visual Dictionary of Architecture. Van Nostrand Reinhold Company. ISBN 0-442-02462-2, p. 266.

[4] Mark Jarzombek, "Pilaster Play" (http://web.mit.edu/mmj4/www/downloads/thresholds28.pdf), *Thresholds* **28 (Winter 2005)**: 34–41,

- Lewis, Philippa and Gillian Darley, *Dictionary of Ornament* (1986) NY: Pantheon

Gallery

Pilaster and capital, Frankfurt, Germany

Puerta de la colegiata, Lerma, Spain

Brick pilasters in English domestic architecture c1650. Princes Risborough Manor House

Beaux-Arts architecture

Beaux-Arts architecture[a][b]
(English: /(ˌ)boʊˈzɑr/, French: [bozaʁ]) expresses the academic neoclassical architectural style taught at the École des Beaux-Arts in Paris. The *style* "Beaux Arts" is above all the cumulative product of two-and-a-half centuries of instruction under the authority, first, of the Académie royale d'architecture (1671–1793), then, following the French Revolution of the late 18th century, of the Architecture section of the Académie des Beaux-Arts (1795–). The organization under the Ancien Régime of the competition for the Grand Prix de Rome in architecture, offering a chance to study in Rome, imprinted its codes and aesthetic on

Palais Garnier (opened 1875) is a cornerpiece of Beaux Arts

the course of instruction, which culminated during the Second Empire (1850–1870) and the Third Republic that followed. The style of instruction that produced Beaux-Arts architecture continued without major interruption until 1968.[1]

The Beaux-Arts style heavily influenced US architecture in the period from 1880 to 1920.[2] Non-French European architects of the period 1860–1914 tended to gravitate toward their own national academic centers rather than fixating on Paris. British architects of Imperial classicism, in a development culminating in Sir Edwin Lutyens's New Delhi government buildings, followed a somewhat more independent course, owing to the cultural politics of the late 19th century.

Training

The Beaux-Arts training emphasized the mainstream examples of Imperial Roman architecture between Augustus and the Severan emperors, Italian Renaissance, and French and Italian Baroque models especially, but the training could then be applied to a broader range of models: Quattrocento Florentine palace fronts or French late Gothic. American architects of the Beaux-Arts generation often returned to Greek models, which had a strong local history in the American Greek Revival of the early 19th century. For the first time, repertories of photographs supplemented meticulous scale drawings and on-site renderings of details.

The last major American building constructed in the Beaux-Arts style, the San Francisco War Memorial Opera House, completed 1932

Some aspects of Beaux-Arts approach could degenerate into mannerisms. Beaux-Arts training made great use of *agrafes*, clasps that links one architectural detail to another; to interpenetration of forms, a Baroque habit; to "speaking architecture" (*architecture parlante*) in which supposed appropriateness of symbolism could be taken to literal-minded extremes.

Beaux-Arts training emphasized the production of quick conceptual sketches, highly-finished perspective presentation drawings, close attention to the program, and knowledgeable detailing. Site considerations tended toward social and urbane contexts.[3]

All architects-in-training passed through the obligatory stages — studying antique models, constructing *analos*, analyses reproducing Greek or Roman models, "pocket" studies and other conventional steps — in the long competition for the few desirable places at the Académie de France à Rome (housed in the Villa Medici) with traditional requirements of sending at intervals the presentation drawings called *envois de Rome*.

Characteristics

Beaux-Arts architecture depended on sculptural decoration along conservative modern lines, employing French and Italian Baroque and Rococo formulas combined with an impressionistic finish and realism. In the façade shown to the right, Diana grasps the cornice she sits on in a natural action typical of Beaux-Arts integration of sculpture with architecture.

Beaux-Arts building decoration presenting images of the Roman goddesses Pomona and Diana. Note the naturalism of the postures and the channeled rustication of the stonework.

Slightly overscaled details, bold scuptural supporting consoles, rich deep cornices, swags and sculptural enrichments in the most bravura finish the client could afford gave employment to several generations of architectural modellers and carvers of Italian and Central European backgrounds. A sense of appropriate idiom at the craftsman level supported the design teams of the first truly modern architectural offices.

Though the Beaux-Arts style embodies an approach to a regenerated spirit within the grand traditions rather than a set of motifs, principal characteristics of Beaux-Arts architecture included:

- Flat roof[2]
- Rusticated and raised first story[2]
- Hierarchy of spaces, from "noble spaces"—grand entrances and staircases— to utilitarian ones
- Arched windows[2]
- Arched and pedimented doors[2]
- Classical details:[2] references to a synthesis of historicist styles and a tendency to eclecticism; fluently in a number of "manners"
- Symmetry[2]
- Statuary,[2] sculpture (bas-relief panels, figural sculptures, sculptural groups), murals, mosaics, and other artwork, all coordinated in theme to assert the identity of the building
- Classical architectural details:[2] balustrades, pilasters, garlands, cartouches, with a prominent display of richly detailed clasps (*agrafes*), brackets and supporting consoles
- Subtle polychromy

Alternating male and female mascarons decorate keystones on the San Francisco City Hall

History

On the eve of World War I the Beaux-Arts style began to find major competitors among the architects of Modernism and the nascent International Style. The prestige of the École gave the *style* "Beaux-Arts" a second wind in combining the new manner with the traditional training.

Beaux-Arts in France

Parisian buildings in the Beaux-Arts style

- École des Beaux-Arts
- LeFuel wings of the Louvre
- Opéra Garnier
- Palais du Trocadéro
- Gare d'Orsay
- Grand Palais, Petit Palais, and the Pont Alexandre III
- Palais de Chaillot.

Pont Alexandre III and Grand Palais in Paris.

Beaux-Arts in the United States

The first American architect to attend the École des Beaux-Arts was Richard Morris Hunt, followed by Henry Hobson Richardson. They were followed by an entire generation. Henry Hobson Richardson absorbed Beaux-Arts lessons in massing and spatial planning, then applied them to Romanesque architectural models that were not characteristic of the Beaux-Arts repertory. His Beaux-Arts training taught him to transcend slavish copying and recreate in the essential fully digested and idiomatic manner of his models. Richardson evolved a highly personal style (Richardsonian Romanesque) freed of historicism that was influential in early Modernism.[4]

Grand Central Terminal (Station opened 1871, Terminal 1903), New York City

The "White City" of the World's Columbian Exposition of 1893 in Chicago was a triumph of the movement and a major impetus for the short-lived City Beautiful movement in the United States. Beaux-Arts city planning, with its Baroque insistence on vistas punctuated by symmetry, eye-catching monuments, axial avenues, uniform cornice heights, a harmonious "ensemble," and a somewhat theatrical nobility and accessible charm, embraced ideals that the ensuing Modernist movement decried or just dismissed.[5] The first US university to institute a Beaux-Arts curriculum is MIT in 1893, when the French architect Constant-Désiré Despradelles was brought to MIT to teach. The Beaux-Arts curriculum was subsequently begun at Columbia University, The University of Pennsylvania, and elsewhere.[6] From 1916, the Beaux-Arts Institute of Design in New York City schooled architects, painters, and sculptors to work as active collaborators.

The best-known architectural firm specializing in Beaux-Arts style was McKim, Mead, and White[7] Among universities designed in the Beaux-Arts style there are, the most notable ones are Columbia University, (commissioned in 1896), designed by McKim, Mead, and White; the University of California, Berkeley (commissioned in 1898), designed by John Galen Howard; the campus of MIT (commissioned in 1913), designed by William W. Bosworth, Carnegie Mellon University (commissioned in 1904), designed by Henry Hornbostel; and the University of Texas (commissioned in 1931), designed by Paul Philippe Cret.

The Iowa, a Beaux-Arts condominium in Washington, D.C.

Though Beaux-Arts architecture of the twentieth century might on its surface appear out of touch with the modern age, steel-frame construction and other modern innovations in engineering techniques and materials were often embraced, as in the 1914–1916 construction of the Carolands Chateau south of San Francisco (which was built with a consciousness of the devastating 1906 earthquake). The noted Spanish structural engineer Rafael Guastavino (1842–1908), famous for his vaultings, known as Guastavino tile work, designed vaults in dozens of Beaux-Arts buildings in the Boston, New York, and elsewhere. Beaux-Arts architecture also brought a civic face to the railroad. (Chicago's Union Station and Detroit's Michigan Central Station are famous American examples of this style.) Two of the best American examples of the Beaux-Arts tradition stand within a few blocks of each other: Grand Central Terminal and the New York Public Library.

American architects working in the Beaux-Arts style

The following individuals, students of the Ecole Des Beaux-Arts, are identified as creating work characteristic of the Beaux-Arts style within the United States:

Andrew Mellon Building

- Otto Eugene Adams
- William A. Boring
- William W. Bosworth
- Arthur Brown Jr
- Daniel Burnham
- Carrère and Hastings
- Paul Philippe Cret
- Ernest Flagg
- C. P. H. Gilbert
- Cass Gilbert
- Thomas Hastings
- Raymond Hood
- Henry Hornbostel
- Richard Morris Hunt
- Charles Klauder
- William Rutherford Mead
- Julia Morgan
- Charles Follen McKim
- Henry Orth
- John Russell Pope
- Henry Hobson Richardson
- Edward Lippincott Tilton
- Horace Trumbauer
- Enoch Hill Turnock
- Stanford White
- Willis Polk

Beaux-Arts in Canada

Beaux-Arts was very prominent in public buildings in Canada in the early 20th Century. Notably all three prairie provinces' legislative buildings are in this style.

Canadian architecture in the Beaux-Arts style

- The NHL sponsored Hockey Hall of Fame (formerly a branch of the Bank of Montreal), Toronto (1885)
- London and Lancashire Life Building, Montreal (1898)
- Old Montreal Stock Exchange Building (1903)
- Royal Alexandra Theatre, Toronto (1906)
- Montreal Museum of Fine Arts, Montreal (1912)
- Government Conference Centre, Ottawa (originally a railway station by Ross and Macdonald, (1912)
- Saskatchewan Legislative Building, Regina (1912)
- Montreal Museum of Fine Arts, 1912
- Alberta Legislative Building, Edmonton (1913)
- Manitoba Legislative Building, Winnipeg, (1920)
- Millennium Centre, Winnipeg, (1920)
- Commemorative Arch, Royal Military College of Canada, in Kingston, Ontario (1923)
- Bank of Nova Scotia, Ottawa (1923–24)
- Union Station, Toronto (1913–27)
- Dominion Square Building, Montreal (1930)
- Canada Life Building, Toronto (1931)
- Sun Life Building, Montreal (1913–1931)
- Supreme Court of Canada Building, Ottawa (1938–1946)

Government Conference Centre, Ottawa

Manitoba Legislative Building, Winnipeg

Canadian architects working in the Beaux-Arts style

- William Sutherland Maxwell
- John M. Lyle
- Ross and Macdonald
- Sproatt & Rolph
- Pearson and Darling
- Ernest Cormier

The Great Hall of Toronto's Union Station

Beaux-Arts in Australia

Both Sydney and Melbourne have some significant examples of the style, where it was typically applied to large solid looking public office buildings and banks during the 1920s.

Port Authority building in Melbourne

- National Theatre, Melbourne (1920)
- General Post Office building, Forrest Place, Perth (1923)
- Argus Building. LaTrobe Street, Melbourne (1927)
- Emily McPherson College of Domestic Economy, Melbourne (1927)
- Commonwealth Bank, Martin Place, Sydney (1928)
- Westpac Bank Building, Elizabeth Street, Brisbane (1928)
- Port Authority building, Melbourne (1928)
- Former Mail Exchange Building, Melbourne
- Herald Weekly Times Building. Flinders Street, Melbourne
- Commonwealth Bank building, Forrest Place, Perth (1933)

Notes

^ **a:** The phrase *Beaux Arts* is usually translated as "Fine Arts" in non-architectural English contexts.

References

[1] Robin Middleton, Editor. *The Beaux-Arts and Nineteenth-century French Architecture.* (London: Thames and Hudson, 1982).

[2] Klein and Fogle, *Clues to American Architecture*, 1986, p.38, ISBN 0-913515-18-3.

[3] Arthur Drexler, Editor, *The Architecture of the École des beaux-arts.* (New York: Museum of Modern Art, 1977).

[4] James Philip Noffsinger. *The Influence of the École des Beaux-arts on the Architects of the United States* (Washington DC., Catholic University of America Press, 1955).

[5] Chafee, Richard. *The Architecture of the École des Beaux-Arts.* New York: Museum of Modern Art, 1977.

[6] Mark Jarzombek. *Designing MIT: Bosworth's New Tech.* Northeastern University Press, 2004.

[7] Richard Guy Wilson. *McKim, Mead & White, Architects* (New York: Rizzoli, 1983)

Further reading

- Reed, Henry Hope and Edmund V. Gillon Jr. 1988. *Beaux-Arts Architecture in New York: A Photographic Guide* (Dover Publications: Mineola NY)

- United States. Commission of Fine Arts. 1978, 1988 (2 vols). *Sixteenth Street Architecture* (The Commission of Fine Arts: Washington, D.C. : The Commission) - profiles of Beaux-Arts architecture in Washington D.C. SuDoc FA 1.2: AR 2.

External links

- New York architecture images, Beaux-Arts gallery (http://www.nyc-architecture.com/STYLES/ STY-BeauxArts.htm)
- Advertisement film about the usage of the Beaux Arts style as a reference in kitchen design (http://www. siematic.com/INT/en/our-kitchens/collection/new-in-the-programme/beauxarts/film/film.html)
- Hallidie Building (http://www.greatbuildings.com/buildings/Hallidie_Building.html)
- Beaux-Arts architecture Buildings/Structures (http://www.ranker.com/list/ beaux-arts-architecture-buildings-and-structures/reference)

George Rogers Clark Memorial Bridge

George Rogers Clark Memorial Bridge	
The George Rogers Clark Memorial Bridge as seen from Louisville Waterfront Park	
Carries	4 lanes of US Route 31
Crosses	Ohio River
Locale	Louisville, Kentucky and Jeffersonville, Indiana
Design	Cantilever bridge
Total length	5746.5 ft (1751.5 m)
Width	38.0 ft (11.6 m)
Longest span	819.6 ft (249.8 m)
Opened	1929
Coordinates	38°15′49″N 85°45′05″W

The **George Rogers Clark Memorial Bridge** is a four-lane cantilevered truss bridge crossing the Ohio River between Louisville, Kentucky and Jeffersonville, Indiana, that carries US 31. It is known locally as the Second Street Bridge.

History

It was designed by Ralph Modjeski and Frank Masters with architectural details handled by Paul Philippe Cret of Philadelphia, and construction began in June 1928 by the American Bridge Company of Pittsburgh at a cost of $4.7 million. It was opened to the public on October 31, 1929 as the **Louisville Municipal Bridge** and operated as a toll bridge. The toll was 35 cents until December 31, 1936, when it became a quarter. The last of the bonds that financed the construction were redeemed in 1946, and the tolls were removed.

Dedication plaque on the bridge

In 1949, the bridge was renamed in honor of George Rogers Clark, recognized as the founder of Louisville.[1]

The bridge was rehabilitated in 1958.

There was a movement in the 1950s to restore tolls, as traffic on the bridge had reached capacity and funding was needed for an additional bridge, but a toll was opposed strongly by most residents. Ultimately most of the funds for two additional bridges (for motor vehicles only) that carry interstate highways came from the federal government.

The Clark Memorial Bridge during Thunder Over Louisville

It was placed on the National Register of Historic Places on March 8, 1984.

In June 2010, Kentucky Governor Steve Beshear and Louisville Mayor Jerry Abramson announced a new $3 million streetscape improvement project directly underneath the Clark Memorial Bridge, a three-block area from Main Street to River Road, which will be transformed into a plaza. This includes a new decorative lighting system under the refurbished Clark Memorial Bridge, wide sidewalks, seats, new pedestrian and festival areas, and extensive plantings, making this an inviting promenade for the new KFC YUM! Center. The project was completed in time for the October 2010 opening of the arena.[2]

Culture

Locally, the Clark Bridge is also known as the **Second Street Bridge**, as Louisville's Second Street leads directly to the bridge. This has never been a formal name, however. There is a pedestrian sidewalk on each side of the bridge deck. The Clark Bridge is the only regional Ohio River bridge currently open to non-motorized traffic.

Since 1991, the bridge has been used as "ground zero" for the annual Thunder Over Louisville event, when a waterfall of fireworks flows along the entire length of the bridge during the fireworks show. This involves traffic being closed for much of the week. This is criticized as it cuts off both the only non-interstate and the only pedestrian route between Louisville and southern Indiana, which can impact local businesses such as bicycle couriers.

The bridge is featured in a scene from the 1981 movie *Stripes* in which Bill Murray drives his cab to the middle of the span, gets out of the vehicle and then tosses his keys into the river below.[3]

See also

- List of crossings of the Ohio River

Notes

[1] Luhan. pp. 105.
[2] "Second Street Transformation to Occur Near arena" (http://www.louisvilleky.gov/economicdevelopment/News/2010/
    TransformationonSecondStreettoTakePlace.htm). .
[3] "Second Street to Third Street" (http://louisville.about.com/od/communityandneighborhoods/ss/MainStTour_4.htm). . Retrieved
    2007-02-22.

References

- Allgeier, M.A. (1983). *Louisville Municipal Bridge, Pylons, and Administrative Building* (http://pdfhost.focus. nps.gov/docs/NRHP/Text/84001578.pdf). Louisville, KY: Louisville Landmarks Commission.
- "Automobile Bridges" (1 ed.). 2001.
- Luhan, Gregory A. (2004). *Louisville Guide* (http://books.google.com/books?id=fai4OHydKKIC). Princeton Architectural Press. ISBN 1568984510.

External links

- George Rogers Clark Memorial Bridge at KentuckyRoads.com (http://www.kentuckyroads.com/images/ clark_bridge/) (unofficial)
- Louisville Art Deco page on Municipal Bridge Building and Pylons (http://www.louisvilleartdeco.com/ architecture/MunicipalBridge/MunicipalBridge.html)

Article Sources and Contributors

Louisville and Nashville Railroad Office Building *Source*: http://en.wikipedia.org/w/index.php?title=Louisville_and_Nashville_Railroad_Office_Building *Contributors*: Appraiser, Bedford,
Biruitorul, Doncram, Einbierbitte, Emeraude, Giraffedata, Neutrality, Ost316, Rividian, Slambo, 3 anonymous edits

Downtown Louisville *Source*: http://en.wikipedia.org/w/index.php?title=Downtown_Louisville *Contributors*: Avala, Bedford, Brando03, Brewcrewer, Censusdata, Choster, Chris the speller,
Chris24, Dale Arnett, Eastlaw, Eggsofamerica, GeckoRoamin, GeorgHH, Jcembree, Kassie, Kbdank71, LanternLight, Lightmouse, LilHelpa, Macewindu239, Maloneyo, Marcika, Mihirgk,
Mindmatrix, Miss Communication, Mucraigdaddy, Ntomasse, Ost316, PeterJohnson, Plasticspork, Plastikspork, Retired username, Rividian, Sashford, Seicer, Spangineer, Stevietheman,
Sylfred1977, Taors, Tlholt02, Traxs7, William Avery, Xin, Xnatedawgx, 77 anonymous edits

Kentucky *Source*: http://en.wikipedia.org/w/index.php?title=Kentucky *Contributors*: 10.26, 16@r, 170.20.175.xxx, 1exec1, 1oddbins1, 2help, 4twenty42o, 55david, 88AdolfLover88, A2Kafir,
A3r0, ARMYxTyler, AVand, Abductive, Acdixon, Addionne, Addison0426, Addshore, Adrian.benko, After Midnight, Agrant131211, Agriculture, Ahassan05, Ahoerstemeier, Airplaneman,
Aitias, Aivazovsky, Ajenchel, Al Silonov, Alan Canon, Alansohn, Alexa411, AlexiusHoratius, Alexwcovington, Algebra101, Allstarecho, AllyUnion, Alpha 40475, AlphaEta, Amazingaro,
Amazonien, American2, Amikeco, Analogue Kid, Anaraug, Anarchy2190, Andy M. Wang, Andy Marchbanks, Andy120290, Andyjsmith, Animum, Anna Lincoln, Anonymous555498,
Antandrus, Anthony, Apusc, Arakunem, Asa Miller, Assed206, Astuishin, Atlantia, Aztom2, B, BD2412, BMetts, BSveen, BabuBhatt, Barek, BarrelProof, Bassbonerocks, Beard rrp, Beland,
Belligero, BenBaker, Bender235, Benhealy, Benjh40, Big iron, Billyjoekoepsel, Bit72101, Bje2089, Bkell, Bkessler23, Bkonrad, Blaise170, Blehfu, Bloo Aardvark, BlueAg09, BlueAzure,
BlueMoonlet, Bluedenim, Bmpeers, Bms4880, Bobblewik, Bobbo, Bobbyllama, Bobo192, Bolivian Unicyclist, Bomac, Bongwarrior, Boom 2010, Borderer, BoycottKY, Bradlydollar,
BrainMarble, Brando03, Brandon, Brbigam, Brian Schlosser42, Brighterorange, Brutannica, Bryan Derksen, BryanG, Bryanwillett, Bsmeathers, Buaidh, Bullshark44, C.Fred, CJLL Wright, CQ,
CSWarren, Caiaffa, CalicoCatLover, Caliga10, Camarcri000, CambridgeBayWeather, Campaigner444, CanisRufus, Canterbury Tail, Caponer, Capricorn42, Carlaude, Casper2k3, Cburnett,
Cchow2, Censusdata, CesarB, Cfrydj, Chanheigeorge, CharlotteWebb, ChocolateSaltyBalls, Choichy5, ChongDae, Chris the speller, Chris24, ChrisRuvolo, Chrislk02, Christian145678,
Chun-hian, Cinik, Circeus, Civil Engineer III, Ckatz, ClairSamoht, Clearlyme, Cliffb, Cmadler, Coasterlover1994, Codydrew, Coemgenus, Coine in the pocket, Cometstyles, CommonsDelinker,
Conankudo51, Confiteordeo, Conversion script, Cool Blue, Cornishgame, Cougar11, Courcelles, CrazyC83, Crobertson, Crunchy Numbers, Cwiki, CylonCAG, Cyndi63, D6, DARTH SIDIOUS
2, DCEdwards1966, DE, DECOR8Rgirl, DJBullfish, DVD R W, Da Packman, Dabomb87, Dadude3320, Dale Arnett, Daniel Case, Danny, DarkFalls, Darkness Productions, Dave, Daven200520,
David.Gaya, DavidHarden, Dawn Bard, Day44581, Dblevins2, Dcolemire, Dcornwall, Dcsohl, Debresser, Deconstructhis, Deflective, Delldot, Deltabeignet, Derek.cashman, Deville,
Dietcokementos, Disavian, Dkdantastic, Don de la Muncha, Doniago, Doom2000, Doulos Christos, Douschonmandick, Download, Dp462090, Dr. Blofeld, Drano, Drclay88, Drumex, Dtox11,
Duckwing, Dweller, Dwo, Dysprosia, ERcheck, ESkog, Eastcote, Eastfrisian, Eco84, Ed g2s, EdBever, Edison, Edivorce, Edward, Ejc891, Ejrubio, ElKevbo, Eldude611, Elikos91, Ellywa,
Elm-39, Emhoo, Enauspeaker, EncMstr, Encephalon, EncycloPetey, Engineer Bob, Epbr123, Epolk, ErgoSum88, Eric, Ericnease, EurekaLott, Everyking, Evice, Evrim98, Ewen, Excirial,
Explicit, FPAtl, Faradayplank, Fartman298, Figgie123, Finalius, Fishing, Fizzleshizzle, FlatFork, Flatterworld, Fledgist, FloNight, Floridahistory, Fogherty V. Tatin, Folic Acid, Footballfan190,
Footwarrior, Fourthords, Fradd, Francs2000, FreplySpang, Funandtrvl, Funnyhat, Furado, G.Freeman123, Gaius Cornelius, Gatita09, Gator87, GeorgHH, Gerry D, Gfoley4, Gilliam, Gimboid13,
Gimmetrow, Giraffedata, Glass Sword, GlassCobra, Gmaxwell, Gogo Dodo, Gold heart, Good Olfactory, GraemeMcRae, Graham87, GregAsche, GregU, Greswik, Gsmgm, Gtbob12, Gurchzilla,
Gurt Posh, Gymnast, Gökhan, Haakon, Had22, Hafrolocks, HalfShadow, HamburgerRadio, Hamtechperson, Happyfatman021, Haymaker, Heegoop, Hellomynameisjoswai, Hellosandimas,
Hendo1270, Heuertag, HexaChord, HiB2Bornot2B, Hmains, Hobartimus, Hoke001, Hottentot, Hu12, Huntington, I dream of horses, IFCAR, IRP, Ianwhite21, Icedman42, Immunize, Imroy,
Incornsyucopia, Instinct, Iridescent, IronChris, Ironholds, It's-is-not-a-genitive, Itascapark, Itfc+canes=me, Itinerant1, Ivirivi00, Ixfd64, J.Steinbock, J.delanoy, J654567, JCO312, JHunterJ,
JMRG, JNRVM, JS Nelson, JW1805, Ja 62, Jaardon, Jack Cox, Jack Merridew, Jacob.jose, Jajhill, Jaknouse, James Byrne69, James Nicol, JamesBWatson, January, Jatkins, Jauhienij, Jaw959,
Jaysing103, Jbamb, Jbfwildcat, Jcembree, Jcravens42, Jduckeck, JdwNYC, Jeffrey O. Gustafson, Jengod, Jhendin, Jiang, Jill Orly, JimIrwin, JimVC3, JimWae, Jimp, JinJian, Jm0371,
Jmeisenhelder, John K, John Reaves, John254, JohnInDC, JohnOwens, Jojhutton, Joncnunn, Jose77, Joseph Dwayne, JoshEdgar, Joyous!, Jspivak86, Jsum, Jt 200075, Juliancolton, Jultemplet,
Junglecat, Juniebjone, Jusdafax, Just Say No, Juzeris, KGasso, Kahuna207, Kahuroa, Kaos456, Karam.Anthony.K, Katalaveno, Katherine, Katieh5584, Kct124, Kd4nuh, Keegan, Keilana, Keith
Edkins, KeithH, Kem0106, Ken g6, Kenaldinho10, Kentucky1333, Kentuckyliberation, Kevin B12, Kevin Myers, KevinI85, Kguiver, Khatru2, Kinneyboy90, Kinston eagle, Kintetsubuffalo,
Kittysrfun, Kleinklein, Knowledge Seeker, KnowledgeOfSelf, Kourc, Kooldude28, Kozuch, Kpjas, Kralahome, Kralizec!, KramerNL, Kronos o, Kumioko, KurtFaulkner, Kwamikagami, Kyace,
Kylepotter, KyraVixen, Kyshooter, L Kensington, L.cash.m, La goutte de pluie, LaOtto, Lamontacranston, Landroo, Lankiveil, Law, LeaveSleaves, Lebanonman19, Ledenierhomme,
Lemonflash, Leo1410, LeoDV, Les Invisibles, Letcher Mountaineer, Leuko, Levelistchampion, Levi93, Levineps, Lightmouse, Lilac Soul, Ling.Nut, LizardJr8, Lizmichael, Logan,
LonghornsU10, Lord Emsworth, Lord Voldemort, Louisvillian, Lreid0975, Lucidwave, Lucky Mitch, Luna Santin, M C Y 1008, MER-C, MJCdetroit, MONGO, MPF, MPerel, Macarenses, Mad
Lothairian, Maddie!, Madisonhenry46, Maelnuneb, Malathos, Malcolm Farmer, Mamyles, Mandall 32, Mandarax, Marek69, Marketinggal, Marmote, Martin451, Marysunshine, Massjit,
Mastrchf91, Matijap, MattTM, MauriceHall, Mav, McCorrection, McSly, Mdmorgan, Meaghan, Meelar, Meepmeepmany, Melsaran, Meltonkt, Mhudson2, Michael Devore, Michael Simpson,
Mike s, MikeWilson, Mikeo, Mikevegas40, Mindcry, Minimac's Clone, Mintleaf, Miskwito, Mister Alcohol, Mitchhelper, Modulatum, Moreau36, Moverton, MrBleu, MrFish, Mu Cow, Mufka,
Mulad, Murraytest, Mwanner, Mxn, Myfatchildisscary, Myweenieiseenormus, N Vale, N734LQ, NCDane, NE2, NEICenergy, NJA, Nakon, Nascar1996, Nasnema, NawlinWiki, Neelix,
NellieBly, Netparrot, Neutrality, Nick Comber, Nick Number, Nihiltres, Ninly, Nirvana23m, Nivek1385, Nivix, Nnemo, Nneonneo, No such user, Noraft, North Shoreman, Nskillen, Ntsimp,
Num1scot, Nuno Tavares, OOODDD, OPMaster, Oda Mari, Ohnoitsjamie, OlEnglish, Old Abe, OmeiKida, Ommnomnomgulp, Omnicog, Omygawsh, Ondewelle, Oroso, Orphan Wiki, Ost316,
OutrunA1C, OverMyHead, OverlordQ, OwenX, Oxymoron83, Ozi, PBP, PFHLai, PL290, PMDrive1061, Pan Dan, Pankkake, Pantallica megadeth, Parkwells, Passionless, Patrick, Patrickneil,
Patricknoddy, Patronik, Pccromeo, Pearle, Peets, Pejman47, Pennst110, Penthrift, Perrytwinkle, Persian Poet Gal, Petrb, Petri Krohn, Pfly, Pgk, Phase Theory, Philip Trueman, Phox6469, Piano
non troppo, Pigman, Pinethicket, Plasticup, Pmac21, Poccil, Pollinator, Polymerbringer, Possum, Postdlf, Prodego, Proficient, Pshent, Psthi, Pvmoutside, Qqqqqq, Quadell, Quantpole, Queson,
QuixoticLife, Qwyrxian, R'n'B, R-HIT, R.T.Gellar, RL0919, Ragib, Railer 103, RainbowOfLight, Rakista, Ramrod1001, Randomfrenchie, Raprat0, Rarefly45, Ratsrule, Ravedave, Rayne GS,
Razorflame, Rd232, Rdikeman, Realm of Shadows, Reaper Eternal, Recognizance, Red Director, Red dwarf, RedCoat1510, Redsully, Reesh, Regoarrarr, Rentman, Retired username, Revas,
RexNL, Reywas92, Rhatsa26X, Rhopkins8, Rich Farmbrough, Richard Keatinge, RichardF, Rick Block, RickK, Rividian, Rjensen, Rjwilmsi, Rlquall, Rmatsu888, Robert LeBlanc, Robomanx,
RobyWayne, Rod Foster, Roger231, Roger2518, Rogerd, Romanm, Romannumeral, Romarin, Root Beers, RoxyTheWolf, Royar, Rreagan007, Rtshawnee, Ruhrfisch, Rumping, RussBlau,
S.ross5, S91by, Sam Dempsey, Sam Korn, Saopaulo1, Sarmatian07, Saturday, SatyrTN, Sayeth, Scanlan, Sceptre, SchfiftyThree, SchnitzelMannGreek, SchuminWeb, Scjessey, Scootey,
Sct112569, Scythian1, Seb az86556, Seicer, Sessenem Kram, Sfmontyo, Sgr2000, Shadowjams, Shanes, Sharkface217, Shirik, Shlomke, Sierra, Simeon24601, Sinn, Sirrom nodnarb yerffej,
Sjakkalle, Sk8er01, SkerHawx, Skyezx, Smashville, Smith03, Smokytop, Smylezsh, Snowolf, Sol!66, Solipsist, Some jerk on the Internet, SonPraises, Sonpraises, Sophus Bie, Scarabezhad,
Sosasos2001, Southernky, Sparsefarce, Spydrlink, Squigglybee, Sreejithk2000, Stallions2010, Stan Shebs, Startstop123, SteelTownMoFo, Steelers130796, Stelio1234567890, Stepshep,
Steve03Mills, Steven Gazeas, Stevewonder2, Stevietheman, Stevotower, Storkynoob, StuffOfInterest, Stuhacking, Swollib, Syra987, TKD, TUF-KAT, TacoMonkey1, Tamer of hope, Tdadkins1,
Teapot37, Techman224, Tedickey, Teh tennisman, Template namespace initialisation script, TenTech, Tewy, The Epopt, The Punk, The Thing That Should Not Be, The Transhumanist, TheFox,
TheHoosierState89, TheNewPhobia, TheYoungThinker, Thedesolatewest, Thehelpfulone, Thejackyl, Theredhouse7, Thesouthernhistorian45, Thingg, Thunderectum, Thunderpooch, Tide rolls,
Tidying Up, Timneu22, Tobby72, Todd Vierling, Tom Allen, Tom mayfair, Tom.k, TomStike, Tomas417, Tomatosoup97, Tommy2010, Tony1, Totaldismair, Tracer9999, Turtle ball 2, Twaz,
TwinCityIL, Ucucha, Ugur Basak, Uklf, Ulric1313, Ultra X987, Underwoodm, Useight, Utcursch, Uyvsdi, Vbofficial, Velvetron, Vendettax, Victor, Vina, Viriditas, Vranak, Vsmith,
Vzbs34, WHeimbigner, WJBscribe, WOSlinker, WPANI, WadeSimMiser, WalkenIkayne, Walrusfighter, Walton One, Wangi, Wapcaplet, Warbuff 4, Wars, Wayne Slam, We are getting pist
now, Weatherman78, Welshbear61, Weregerbil, Wghornsby, WhisperToMe, Whocares81, WikHead, Wikicrasher, Wikieditor06, WikiIrsc, Wikipedianinthehouse, Wikipelli, Wildcat156, Will
Beback, William Avery, Willking1979, Wknight94, Wmahan, Woohookitty, WxGopher, X96lee15, Xaosflux, Xazbq, Xdenizen, XisTheBest, Xyzzyva, Yaf, Yamamoto Ichiro, Yandman,
Youngamerican, YourNight, Yuckfoo, Zaui, Zntrip, Zoe, Zoicon5, ZooFari, Zzuuzz, Zzyzx11, Δ, 1737 anonymous edits

Louisville and Nashville Railroad *Source*: http://en.wikipedia.org/w/index.php?title=Louisville_and_Nashville_Railroad *Contributors*: Acdixon, Alco83, Astatine-210, Baxterguy, Bedford,
Bhuck, Biscuittin, Carrie.daniels, Chessie train, CommonsDelinker, DanTD, DanielRigal, Dravecky, Drewish, Eastlaw, Edward, Fawcett5, Fyrael, Graham87, Hugo999, Infrogmation, Insanity
Incarnate, Jackieman402, Jdm280, Jeroldc, Jolomo, Lotje, Mackensen, Mangoe, Mark Sublette, MrPrada, NE2, OllieFury, Ost316, Plasma east, Realkyhick, Retired username, RevelationDirect,
RivGuySC, Rlquall, SPUI, Scott Mingus, Siebrand, Slambo, Stevietheman, Sylfred1977, Textorus, Tomas417, Vaoverland, Woodsylass, Xnatedawgx, Zaharous, ZekeMacNeil, Zpb52, 22
anonymous edits

Pilaster *Source*: http://en.wikipedia.org/w/index.php?title=Pilaster *Contributors*: Alan Liefting, Arpingstone, BD2412, Beetstra, Brosi, Centrx, Circeus, Cyberman, DVD R W, Dogears, Dreish,
Elekhh, Guilty@eskimo.com, Hede2000, Historicist, Jdsteakley, Kazubon, Levent, Manuel Anastácio, Martínhache, Mattisse, Mihai Andrei, Mjc42, Notinasnaid, Phil Sawyer, Qrsdogg, R'n'B,
Safalra, Sailko, Sfdan, Tail, Teda13, Tysto, Vanished User 1004, Waysider1925, Wetman, William Avery, Wmpearl, 29 anonymous edits

Beaux-Arts architecture *Source*: http://en.wikipedia.org/w/index.php?title=Beaux-Arts_architecture *Contributors*: A.Roz, AgnosticPreachersKid, Amleth, Andy Marchbanks, AngieMcMaster,
Axeman89, Biatch, Bozzio, Brosi, Brownh2o, Bumtorange72, Carptrash, Conscious, Cornellrockey, Cp98ak, DMG413, Davepape, DavidLevinson, Dddstone, Dogears, DonDaMon, Drmies,
Drphilharmonic, EliasTheHorse, Elkman, Elockid, Equendil, Erianna, Evrik, Fenbaud, FloNight, GCappy, GcSwRhIc, Ghirlandajo, Grosscha, Gurt Posh, Gyrobo, Ham, Hede2000, Heitor CJ,
HooHooH, Irishman259, James Russiello, Jennifer Ames, Jesster79, Jonathan.s.kt, Joopercoopers, Joseph Solis in Australia, Joy, Kablammo, Kent Wang, Kevlar67, Kintetsubuffalo,
Kwamikagami, L Glidewell, Lambiam, Leonard G., Lockley, Lquilter, M@sk, MPS, Mackensen, Magog the Ogre, Marcbela, Mattis, Mcginnly, Michaelyee, Murderbike, NCSUSCRC,
Neddyseagoon, Nyttend, Ousepe, Parkwells, Phy6, Pissant, Polynova, President Rhapsody, R'n'B, Ranjithsutari, Remuel, Retired username, Rich Farmbrough, Rohithd reddy, Saltlakejohn,
Sandover, Saucybetty, Series6studios, Shawn in Montreal, Sillygoose0203, SimonP, Teda13, Valerius Tygart, Vaquero100, Verne Equinox, Victoriaedwards, Wetman, Winterst, Yahel Guhan,
Yassie, ZS, Zigger, Zundark, 84 anonymous edits

George Rogers Clark Memorial Bridge *Source*: http://en.wikipedia.org/w/index.php?title=George_Rogers_Clark_Memorial_Bridge *Contributors*: Analogue Kid, Appraiser, Bedford, Bleakcomb, ChangeAgent, D6, Davidharpe, Dogears, DwightKingsbury, Elkman, GrahamHardy, HarveyHenkelmann, Hmains, Jllm06, John of Reading, NE2, Ost316, Purple hills, Retired username, Stevietheman, Vantey, VerruckteDan, Xnatedawgx, 9 anonymous edits

Image Sources, Licenses and Contributors

File:L&N Office Building Louisville closeup.JPG *Source*: http://en.wikipedia.org/w/index.php?title=File:L&N_Office_Building_Louisville_closeup.JPG *License*: unknown *Contributors*: Original uploader was Bedford at en.wikipedia

file:USA Kentucky location map.svg *Source*: http://en.wikipedia.org/w/index.php?title=File:USA_Kentucky_location_map.svg *License*: unknown *Contributors*: User:Alexrk2

File:Red pog.svg *Source*: http://en.wikipedia.org/w/index.php?title=File:Red_pog.svg *License*: unknown *Contributors*: Anomie

File:Louisville aerial.jpg *Source*: http://en.wikipedia.org/w/index.php?title=File:Louisville_aerial.jpg *License*: unknown *Contributors*: Louisville and Nashville Railroad Company/Family Lines Rail System Archives

Image:Louisville Skyline.jpg *Source*: http://en.wikipedia.org/w/index.php?title=File:Louisville_Skyline.jpg *License*: unknown *Contributors*: Chris Watson

Image:AEGON Center.JPG *Source*: http://en.wikipedia.org/w/index.php?title=File:AEGON_Center.JPG *License*: unknown *Contributors*: Michael Baker

Image:Levy building louisville.jpg *Source*: http://en.wikipedia.org/w/index.php?title=File:Levy_building_louisville.jpg *License*: unknown *Contributors*: User:W.marsh

Image:Columbia-building.jpg *Source*: http://en.wikipedia.org/w/index.php?title=File:Columbia-building.jpg *License*: unknown *Contributors*: USA

Image:The Brown Hotel, Louisville, KY.jpg *Source*: http://en.wikipedia.org/w/index.php?title=File:The_Brown_Hotel,_Louisville,_KY.jpg *License*: unknown *Contributors*: Original uploader was Derek.cashman at en.wikipedia

Image:Heyburn side.jpg *Source*: http://en.wikipedia.org/w/index.php?title=File:Heyburn_side.jpg *License*: unknown *Contributors*: User:W.marsh

Image:Pendennis club c1906.jpg *Source*: http://en.wikipedia.org/w/index.php?title=File:Pendennis_club_c1906.jpg *License*: unknown *Contributors*: Retired username

Image:Louemaindev.jpg *Source*: http://en.wikipedia.org/w/index.php?title=File:Louemaindev.jpg *License*: unknown *Contributors*: Censusdata

Image:Loulibertygreens.jpg *Source*: http://en.wikipedia.org/w/index.php?title=File:Loulibertygreens.jpg *License*: unknown *Contributors*: Censusdata

Image:Louisville Fourthstreetlive.jpg *Source*: http://en.wikipedia.org/w/index.php?title=File:Louisville_Fourthstreetlive.jpg *License*: unknown *Contributors*: Censusdata

Image:Thunder over louisville 2006.jpg *Source*: http://en.wikipedia.org/w/index.php?title=File:Thunder_over_louisville_2006.jpg *License*: unknown *Contributors*: Shawn Skriver

Image:LouisvilleSluggerMusem.jpg *Source*: http://en.wikipedia.org/w/index.php?title=File:LouisvilleSluggerMusem.jpg *License*: unknown *Contributors*: User Derek.cashman on en.wikipedia

Image:Louisville Skatepark-night-2002.jpg *Source*: http://en.wikipedia.org/w/index.php?title=File:Louisville_Skatepark-night-2002.jpg *License*: unknown *Contributors*: Original uploader was Zeroasterisk at en.wikipedia

Image:AliCenter.jpg *Source*: http://en.wikipedia.org/w/index.php?title=File:AliCenter.jpg *License*: unknown *Contributors*: Original uploader was Stevietheman at en.wikipedia

Image:FIHM.jpg *Source*: http://en.wikipedia.org/w/index.php?title=File:FIHM.jpg *License*: unknown *Contributors*: Original uploader was FrazierMuseum at en.wikipedia

Image:LouisvilleGlassworks.jpg *Source*: http://en.wikipedia.org/w/index.php?title=File:LouisvilleGlassworks.jpg *License*: unknown *Contributors*: Bedford

Image:louisville sistercities.jpg *Source*: http://en.wikipedia.org/w/index.php?title=File:Louisville_sistercities.jpg *License*: unknown *Contributors*: AnRo0002, Urban, Xnatedawgx

Image:LouisvilleSkyscrapers.jpg *Source*: http://en.wikipedia.org/w/index.php?title=File:LouisvilleSkyscrapers.jpg *License*: unknown *Contributors*: Original uploader was Stevietheman at en.wikipedia

Image:Aegoncenterlou.jpg *Source*: http://en.wikipedia.org/w/index.php?title=File:Aegoncenterlou.jpg *License*: unknown *Contributors*: Censusdata

Image:Picture 1680.jpg *Source*: http://en.wikipedia.org/w/index.php?title=File:Picture_1680.jpg *License*: unknown *Contributors*: Censusdata

Image:PNC_Plaza.JPG *Source*: http://en.wikipedia.org/w/index.php?title=File:PNC_Plaza.JPG *License*: unknown *Contributors*: User:SimonP

Image:Twintowerslou.jpg *Source*: http://en.wikipedia.org/w/index.php?title=File:Twintowerslou.jpg *License*: unknown *Contributors*: Censusdata

Image:Loumaindev1.jpg *Source*: http://en.wikipedia.org/w/index.php?title=File:Loumaindev1.jpg *License*: unknown *Contributors*: Censusdata

Image:Jewishopstowers.jpg *Source*: http://en.wikipedia.org/w/index.php?title=File:Jewishopstowers.jpg *License*: unknown *Contributors*: Censusdata

Image:Loudwntmotels.jpg *Source*: http://en.wikipedia.org/w/index.php?title=File:Loudwntmotels.jpg *License*: unknown *Contributors*: Censusdata

Image:Ulmedconstr.jpg *Source*: http://en.wikipedia.org/w/index.php?title=File:Ulmedconstr.jpg *License*: unknown *Contributors*: Censusdata

Image:Newmotellou.jpg *Source*: http://en.wikipedia.org/w/index.php?title=File:Newmotellou.jpg *License*: unknown *Contributors*: Censusdata

File:Flag of Kentucky.svg *Source*: http://en.wikipedia.org/w/index.php?title=File:Flag_of_Kentucky.svg *License*: unknown *Contributors*: Commonwealth of Kentucky

File:Seal_of_Kentucky.svg *Source*: http://en.wikipedia.org/w/index.php?title=File:Seal_of_Kentucky.svg *License*: unknown *Contributors*: Commonwealth of Kentucky

File:Kentucky in United States.svg *Source*: http://en.wikipedia.org/w/index.php?title=File:Kentucky_in_United_States.svg *License*: unknown *Contributors*: TUBS

Image:Speakerlink.svg *Source*: http://en.wikipedia.org/w/index.php?title=File:Speakerlink.svg *License*: unknown *Contributors*: Woodstone. Original uploader was Woodstone at en.wikipedia

File:Country road in Kentucky.jpg *Source*: http://en.wikipedia.org/w/index.php?title=File:Country_road_in_Kentucky.jpg *License*: unknown *Contributors*: Original uploader was Censusdata at en.wikipedia. Later version(s) were uploaded by Pd THOR at en.wikipedia.

File:Map of Kentucky NA.png *Source*: http://en.wikipedia.org/w/index.php?title=File:Map_of_Kentucky_NA.png *License*: unknown *Contributors*: Huebi, Jmabel

File:KYphysiography.jpg *Source*: http://en.wikipedia.org/w/index.php?title=File:KYphysiography.jpg *License*: unknown *Contributors*: Original uploader was Lamontacranston at en.wikipedia

File:Khp.jpg *Source*: http://en.wikipedia.org/w/index.php?title=File:Khp.jpg *License*: unknown *Contributors*: Wes Blevins

File:Kybaldcypress.jpg *Source*: http://en.wikipedia.org/w/index.php?title=File:Kybaldcypress.jpg *License*: unknown *Contributors*: User:Southernky

File:View from Pine Mountain (Kentucky).jpg *Source*: http://en.wikipedia.org/w/index.php?title=File:View_from_Pine_Mountain_(Kentucky).jpg *License*: unknown *Contributors*: User:J654567

File:Picture 491.jpg *Source*: http://en.wikipedia.org/w/index.php?title=File:Picture_491.jpg *License*: unknown *Contributors*: Original uploader was Censusdata at en.wikipedia

File:WaterfrontPkDwnt.jpg *Source*: http://en.wikipedia.org/w/index.php?title=File:WaterfrontPkDwnt.jpg *License*: unknown *Contributors*: User:Angry Aspie

File:Half Moon, Kentucky.JPG *Source*: http://en.wikipedia.org/w/index.php?title=File:Half_Moon,_Kentucky.JPG *License*: unknown *Contributors*: Beeblebrox, DragonflySixtyseven, Mrherbalwarrior

File:LincolnBirthplaceJBM82908.jpg *Source*: http://en.wikipedia.org/w/index.php?title=File:LincolnBirthplaceJBM82908.jpg *License*: unknown *Contributors*: Censusdata at en.wikipedia. Later version was cropped by Fourthords at en.wikipedia.

File:Lincoln and Davis Statue.jpg *Source*: http://en.wikipedia.org/w/index.php?title=File:Lincoln_and_Davis_Statue.jpg *License*: unknown *Contributors*: JW1805, 1 anonymous edits

File:U.S. Marine Hospital, Louisville, Kentucky.jpg *Source*: http://en.wikipedia.org/w/index.php?title=File:U.S._Marine_Hospital,_Louisville,_Kentucky.jpg *License*: unknown *Contributors*: User:SreeBot

File:KY Governors Mansion.png *Source*: http://en.wikipedia.org/w/index.php?title=File:KY_Governors_Mansion.png *License*: unknown *Contributors*: Original uploader was Acdixon at en.wikipedia

File:KY State Capitol.jpg *Source*: http://en.wikipedia.org/w/index.php?title=File:KY_State_Capitol.jpg *License*: unknown *Contributors*: w:User:RXUYDCRXUYDC

File:KY-districts-108.JPG *Source*: http://en.wikipedia.org/w/index.php?title=File:KY-districts-108.JPG *License*: unknown *Contributors*: Original uploader was Seth Ilys at en.wikipedia

File:Kentucky population map.png *Source*: http://en.wikipedia.org/w/index.php?title=File:Kentucky_population_map.png *License*: unknown *Contributors*: Original uploader was JimIrwin at en.wikipedia

File:CollegeoftheBible-LexKY.JPG *Source*: http://en.wikipedia.org/w/index.php?title=File:CollegeoftheBible-LexKY.JPG *License*: unknown *Contributors*: unknown

File:Kentucky quarter, reverse side, 2001.jpg *Source*: http://en.wikipedia.org/w/index.php?title=File:Kentucky_quarter,_reverse_side,_2001.jpg *License*: unknown *Contributors*: T. James Ferrell, employee of the US Mint

File:2007-Toyota-Camry-SE.jpg *Source*: http://en.wikipedia.org/w/index.php?title=File:2007-Toyota-Camry-SE.jpg *License*: unknown *Contributors*: IFCAR

File:2004-2007 Ford F-150 Lariat.jpg *Source*: http://en.wikipedia.org/w/index.php?title=File:2004-2007_Ford_F-150_Lariat.jpg *License*: unknown *Contributors*: IFCAR

File:Kentucky.JPG *Source*: http://en.wikipedia.org/w/index.php?title=File:Kentucky.JPG *License*: unknown *Contributors*: User:ErgoSum88

GNU Free Documentation License Version 1.2, November 2002 Copyright (C) 2000,2001,2002 Free Software Foundation, Inc. 59 Temple Place, Suite 330, Boston, MA 02111-1307 USA Everyone is permitted to copy and distribute verbatim copies of this license document, but changing it is not allowed.

0. PREAMBLE

The purpose of this License is to make a manual, textbook, or other functional and useful document "free" in the sense of freedom: to assure everyone the effective freedom to copy and redistribute it, with or without modifying it, either commercially or noncommercially. Secondarily, this License preserves for the author and publisher a way to get credit for their work, while not being considered responsible for modifications made by others. This License is a kind of "copyleft", which means that derivative works of the document must themselves be free in the same sense. It complements the GNU General Public License, which is a copyleft license designed for free software. We have designed this License in order to use it for manuals for free software, because free software needs free documentation: a free program should come with manuals providing the same freedoms that the software does. But this License is not limited to software manuals; it can be used for any textual work, regardless of subject matter or whether it is published as a printed book. We recommend this License principally for works whose purpose is instruction or reference.

1. APPLICABILITY AND DEFINITIONS

This License applies to any manual or other work, in any medium, that contains a notice placed by the copyright holder saying it can be distributed under the terms of this License. Such a notice grants a world-wide, royalty-free license, unlimited in duration, to use that work under the conditions stated herein. The "Document", below, refers to any such manual or work. Any member of the public is a licensee, and is addressed as "you". You accept the license if you copy, modify or distribute the work in a way requiring permission under copyright law. A "Modified Version" of the Document means any work containing the Document or a portion of it, either copied verbatim, or with modifications and/or translated into another language. A "Secondary Section" is a named appendix or a front-matter section of the Document that deals exclusively with the relationship of the publishers or authors of the Document to the Document's overall subject (or to related matters) and contains nothing that could fall directly within that overall subject. (Thus, if the Document is in part a textbook of mathematics, a Secondary Section may not explain any mathematics.) The relationship could be a matter of historical connection with the subject or with related matters, or of legal, commercial, philosophical, ethical or political position regarding them. The "Invariant Sections" are certain Secondary Sections whose titles are designated, as being those of Invariant Sections, in the notice that says that the Document is released under this License. If a section does not fit the above definition of Secondary then it is not allowed to be designated as Invariant. The Document may contain zero Invariant Sections. If the Document does not identify any Invariant Sections then there are none. The "Cover Texts" are certain short passages of text that are listed, as Front-Cover Texts or Back-Cover Texts, in the notice that says that the Document is released under this License. A Front-Cover Text may be at most 5 words, and a Back-Cover Text may be at most 25 words. A "Transparent" copy of the Document means a machine-readable copy, represented in a format whose specification is available to the general public, that is suitable for revising the document straightforwardly with generic text editors or (for images composed of pixels) generic paint programs or (for drawings) some widely available drawing editor, and that is suitable for input to text formatters or for automatic translation to a variety of formats suitable for input to text formatters. A copy made in an otherwise Transparent file format whose markup, or absence of markup, has been arranged to thwart or discourage subsequent modification by readers is not Transparent. An image format is not Transparent if used for any substantial amount of text. A copy that is not "Transparent" is called "Opaque". Examples of suitable formats for Transparent copies include plain ASCII without markup, Texinfo input format, LaTeX input format, SGML or XML using a publicly available DTD, and standard-conforming simple HTML, PostScript or PDF designed for human modification. Examples of transparent image formats include PNG, XCF and JPG. Opaque formats include proprietary formats that can be read and edited only by proprietary word processors, SGML or XML for which the DTD and/or processing tools are not generally available, and the machine-generated HTML, PostScript or PDF produced by some word processors for output purposes only. The "Title Page" means, for a printed book, the title page itself, plus such following pages as are needed to hold, legibly, the material this License requires to appear in the title page. For works in formats which do not have any title page as such, "Title Page" means the text near the most prominent appearance of the work's title, preceding the beginning of the body of the text. A section "Entitled XYZ" means a named subunit of the Document whose title either is precisely XYZ or contains XYZ in parentheses following text that translates XYZ in another language. (Here XYZ stands for a specific section name mentioned below, such as "Acknowledgements", "Dedications", "Endorsements", or "History".) To "Preserve the Title" of such a section when you modify the Document means that it remains a section "Entitled XYZ" according to this definition. The Document may include Warranty Disclaimers next to the notice which states that this License applies to the Document. These Warranty Disclaimers are considered to be included by reference in this License, but only as regards disclaiming warranties: any other implication that these Warranty Disclaimers may have is void and has no effect on the meaning of this License.

2. VERBATIM COPYING

You may copy and distribute the Document in any medium, either commercially or noncommercially, provided that this License, the copyright notices, and the license notice saying this License applies to the Document are reproduced in all copies, and that you add no other conditions whatsoever to those of this License. You may not use technical measures to obstruct or control the reading or further copying of the copies you make or distribute. However, you may accept compensation in exchange for copies. If you distribute a large enough number of copies you must also follow the conditions in section 3. You may also lend copies, under the same conditions stated above, and you may publicly display copies.

3. COPYING IN QUANTITY

If you publish printed copies (or copies in media that commonly have printed covers) of the Document, numbering more than 100, and the Document's license notice requires Cover Texts, you must enclose the copies in covers that carry, clearly and legibly, all these Cover Texts: Front-Cover Texts on the front cover, and Back-Cover Texts on the back cover. Both covers must also clearly and legibly identify you as the publisher of these copies. The front cover must present the full title with all words of the title equally prominent and visible. You may add other material on the covers in addition. Copying with changes limited to the covers, as long as they preserve the title of the Document and satisfy these conditions, can be treated as verbatim copying in other respects. If the required texts for either cover are too voluminous to fit legibly, you should put the first ones listed (as many as fit reasonably) on the actual cover, and continue the rest onto adjacent pages. If you publish or distribute Opaque copies of the Document numbering more than 100, you must either include a machine-readable Transparent copy along with each Opaque copy, or state in or with each Opaque copy a computer-network location from which the general network-using public has access to download using public-standard network protocols a complete Transparent copy of the Document, free of added material. If you use the latter option, you must take reasonably prudent steps, when you begin distribution of Opaque copies in quantity, to ensure that this Transparent copy will remain thus accessible at the stated location until at least one year after the last time you distribute an Opaque copy (directly or through your agents or retailers) of that edition to the public. It is requested, but not required, that you contact the authors of the Document well before redistributing any large number of copies, to give them a chance to provide you with an updated version of the Document.

4. MODIFICATIONS

You may copy and distribute a Modified Version of the Document under the conditions of sections 2 and 3 above, provided that you release the Modified Version under precisely this License, with the Modified Version filling the role of the Document, thus licensing distribution and modification of the Modified Version to whoever possesses a copy of it. In addition, you must do these things in the Modified Version: A. Use in the Title Page (and on the covers, if any) a title distinct from that of the Document, and from those of previous versions (which should, if there were any, be listed in the History section of the Document). You may use the same title as a previous version if the original publisher of that version gives permission. B. List on the Title Page, as authors, one or more persons or entities responsible for authorship of the modifications in the Modified Version, together with at least five of the principal authors of the Document (all of its principal authors, if it has fewer than five), unless they release you from this requirement. C. State on the Title page the name of the publisher of the Modified Version, as the publisher. D. Preserve all the copyright notices of the Document. E. Add an appropriate copyright notice for your modifications adjacent to the other copyright notices. F. Include, immediately after the copyright notices, a license notice giving the public permission to use the Modified Version under the terms of this License, in the form shown in the Addendum below. G. Preserve in that license notice the full lists of Invariant Sections and required Cover Texts given in the Document's license notice. H. Include an unaltered copy of this License. I. Preserve the section Entitled "History", Preserve its Title, and add to it an item stating at least the title, year, new authors, and publisher of the Modified Version as given on the Title Page. If there is no section Entitled "History" in the Document, create one stating the title, year, authors, and publisher of the Document as given on its Title Page, then add an item describing the Modified Version as stated in the previous sentence. J. Preserve the network location, if any, given in the Document for public access to a Transparent copy of the Document, and likewise the network locations given in the Document for previous versions it was based on. These may be placed in the "History" section. You may omit a network location for a work that was published at least four years before the Document itself, or if the original publisher of the version it refers to gives permission. K. For any section Entitled "Acknowledgements" or "Dedications", Preserve the Title of the section, and preserve in the section all the substance and tone of each of the contributor acknowledgements and/or dedications given therein. L. Preserve all the Invariant Sections of the Document, unaltered in their text and in their titles. Section numbers or the equivalent are not considered part of the section titles. M. Delete any section Entitled "Endorsements". Such a section may not be included in the Modified Version. N. Do not retitle any existing section to be Entitled "Endorsements" or to conflict in title with any Invariant Section. O. Preserve any Warranty Disclaimers. If the Modified Version includes new front-matter sections or appendices that qualify as Secondary Sections and contain no material copied from the Document, you may at your option designate some or all of these sections as invariant. To do this, add their titles to the list of Invariant Sections in the Modified Version's license notice. These titles must be distinct from any other section titles. You may add a section Entitled "Endorsements", provided it contains nothing but endorsements of your Modified Version by various parties--for example, statements of peer review or that the text has been approved by an organization as the authoritative definition of a standard. You may add a passage of up to five words as a Front-Cover Text, and a passage of up to 25 words as a Back-Cover Text, to the end of the list of Cover Texts in the Modified Version. Only one passage of Front-Cover Text and one of Back-Cover Text may be added by (or through arrangements made by) any one entity. If the Document already includes a cover text for the same cover, previously added by you or by arrangement made by the same entity you are acting on behalf of, you may not add another; but you may replace the old one, on explicit permission from the previous publisher that added the old one. The author(s) and publisher(s) of the Document do not by this License give permission to use their names for publicity for or to assert or imply endorsement of any Modified Version.

5. COMBINING DOCUMENTS

You may combine the Document with other documents released under this License, under the terms defined in section 4 above for modified versions, provided that you include in the combination all of the Invariant Sections of all of the original documents, unmodified, and list them all as Invariant Sections of your combined work in its license notice, and that you preserve all their Warranty Disclaimers. The combined work need only contain one copy of this License, and multiple identical Invariant Sections may be replaced with a single copy. If there are multiple Invariant Sections with the same name but different contents, make the title of each such section unique by adding at the end of it, in parentheses, the name of the original author or publisher of that section if known, or else a unique number. Make the same adjustment to the section titles in the list of Invariant Sections in the license notice of the combined work. In the combination, you must combine any sections Entitled "History" in the various original documents, forming one section Entitled "History"; likewise combine any sections Entitled "Acknowledgements", and any sections Entitled "Dedications". You must delete all sections Entitled "Endorsements".

6. COLLECTIONS OF DOCUMENTS

You may make a collection consisting of the Document and other documents released under this License, and replace the individual copies of this License in the various documents with a single copy that is included in the collection, provided that you follow the rules of this License for verbatim copying of each of the documents in all other respects. You may extract a single document from such a collection, and distribute it individually under this License, provided you insert a copy of this License into the extracted document, and follow this License in all other respects regarding verbatim copying of that document.

7. AGGREGATION WITH INDEPENDENT WORKS

A compilation of the Document or its derivatives with other separate and independent documents or works, in or on a volume of a storage or distribution medium, is called an "aggregate" if the copyright resulting from the compilation is not used to limit the legal rights of the compilation's users beyond what the individual works permit. When the Document is included in an aggregate, this License does not apply to the other works in the aggregate which are not themselves derivative works of the Document. If the Cover Text requirement of section 3 is applicable to these copies of the Document, then if the Document is less than one half of the entire aggregate, the Document's Cover Texts may be placed on covers that bracket the Document within the aggregate, or the electronic equivalent of covers if the Document is in electronic form. Otherwise they must appear on printed covers that bracket the whole aggregate.

8. TRANSLATION

Translation is considered a kind of modification, so you may distribute translations of the Document under the terms of section 4. Replacing Invariant Sections with translations requires special permission from their copyright holders, but you may include translations of some or all Invariant Sections in addition to the original versions of these Invariant Sections. You may include a translation of this License, and all the license notices in the Document, and any Warranty Disclaimers, provided that you also include the original English version of this License and the original versions of those notices and disclaimers. In case of a disagreement between the translation and the original version of this License or a notice or disclaimer, the original version will prevail. If a section in the Document is Entitled "Acknowledgements", "Dedications", or "History", the requirement (section 4) to Preserve its Title (section 1) will typically require changing the actual title.

9. TERMINATION

You may not copy, modify, sublicense, or distribute the Document except as expressly provided for under this License. Any other attempt to copy, modify, sublicense or distribute the Document is void, and will automatically terminate your rights under this License. However, parties who have received copies, or rights, from you under this License will not have their licenses terminated so long as such parties remain in full compliance.

10. FUTURE REVISIONS OF THIS LICENSE

The Free Software Foundation may publish new, revised versions of the GNU Free Documentation License from time to time. Such new versions will be similar in spirit to the present version, but may differ in detail to address new problems or concerns. See http://www.gnu.org/copyleft/. Each version of the License is given a distinguishing version number. If the Document specifies that a particular numbered version of this License "or any later version" applies to it, you have the option of following the terms and conditions either of that specified version or of any later version that has been published (not as a draft) by the Free Software Foundation. If the Document does not specify a version number of this License, you may choose any version ever published (not as a draft) by the Free Software Foundation. ADDENDUM: How to use this License for your documents To use this License in a document you have written, include a copy of the License in the document and put the following copyright and license notices just after the title page: Copyright (c) YEAR YOUR NAME. Permission is granted to copy, distribute and/or modify this document under the terms of the GNU Free Documentation License, Version 1.2 or any later version published by the Free Software Foundation; with no Invariant Sections, no Front-Cover Texts, and no Back-Cover Texts. A copy of the license is included in the section entitled "GNU Free Documentation License". If you have Invariant Sections, Front-Cover Texts and Back-Cover Texts, replace the "with...Texts." line with this: with the Invariant Sections being LIST THEIR TITLES, with the Front-Cover Texts being LIST, and with the Back-Cover Texts being LIST. If you have Invariant Sections without Cover Texts, or some other combination of the three, merge those two alternatives to suit the situation. If your document contains nontrivial examples of program code, we recommend releasing these examples in parallel under your choice of free software license, such as the GNU General Public License, to permit their use in free software.

Printed by Books on Demand GmbH, Norderstedt / Germany